EXTRACTION OF MINERALS AND ENERGY: TODAY'S DILEMMAS

INTERNATIONAL WORKSHOP ON ENVIRONMENTAL
PROBLEMS OF THE EXTRACTIVE INDUSTRIES,
WRIGHT STATE UNIVERSITY, 1973

EXTRACTION OF MINERALS
AND ENERGY:
TODAY'S DILEMMAS

RAUL A. DEJU, PhD

Department of Geology
Wright State University
Dayton, Ohio

ann arbor science PUBLISHERS INC.
POST OFFICE BOX 1425 • ANN ARBOR, MICH. 48106

Papers in this volume were presented at
the International Workshop on Environ-
mental Problems of the Extractive
Industries jointly sponsored by the
Geological Society of America and the
Society of Mining Engineers of AIME.

Dinosaurs with their small
brains were able to survive
millions of years. . . .
Will man be around during
the twenty-first century?

PREFACE

As I write the preface to this book I am being
exposed to the "energy crisis" from all angles:
gasoline is more expensive, shortages of fuels are
developing, politicians are concerned, and the news
media carry alarming headlines.

In September, 1972, (before the "energy hula-
baloo" really got going), I initiated, with the full
support of the Geological Society of America (GSA)
and the Society of Mining Engineers of AIME (SME),
the plans to hold an International Workshop on the
Environmental Problems of the Extractive Industries.
The Workshop was successfully held at Wright State
University in Dayton, Ohio, June 10-13, 1973 with
the support of GSA, SME, and several state and local
organizations including the Water Management Association of Ohio. Topics covered at the workshop
included:

- an appraisal of our present energy, minerals,
  and water resource base
- an evaluation of the water and air pollution
  problems directly attributable to the extractive
  industries
- a study of the waste disposal and land reclamation problems caused by the petroleum and mining
  industries
- an in-depth analysis of the economics of
  pollution control in the mining and petroleum
  industries, and the attitudes of individual
  citizens toward pollution control in these
  industries.

This volume has been prepared in order to make
the main papers presented at the workshop on such
timely topics available to the largest possible
audience. Papers have been grouped into sections
and each section is preceded by introductory comments

vii

outlining the scope of the papers included. It is
hoped that the papers will give the reader a better
insight into the main environmental problems plaguing
the extractive industries today. It is also hoped
that such problems will challenge young people to
spend their time, to study, and to seek solutions.

As is the case with any multidisciplinary
endeavor, many people deserve thanks. First, I
would like to acknowledge the support of the authors
who presented the papers and the many attendees to
the workshop. Without their help this volume would
not have been possible. Thanks are also due to Dr.
Edwin Eckel of GSA, Mr. John C. Fox of SME, L.
Bennett Coy of the Water Management Association of
Ohio and my colleagues in the faculty, staff, and
student body of Wright State University who worked
to make this project successful.

Dr. Raul A. Deju
Coordinator, International
  Workshop, Environmental
  Problems of the Extrac-
  tive Industries
Dayton, Ohio
January 1974

CONTENTS

SECTION IV. RESOURCES OUT OF PLACE

SECTION V. SOCIAL FACTORS

SECTION I

RESOURCES

1. INTRODUCTION

In the past few years the growth in consumption
of energy coupled with political factors has led the
United States head on into a full-fledged energy
crisis. In this section we will take a critical
look at our present position in relation to energy
sources, mineral resources, and water. An accurate
appraisal of supply and demand of these valuable
commodities is urgently needed in order to avoid
future crises caused by the lack of any one commodity.

SOURCES OF ENERGY

Our main sources of energy today are petroleum
and natural gas. These two fossil fuels are the key
to our industrial progress. Although the United
States comprises only 6% of the world population,
it consumes one-third of the world's energy. If we
divide the world into the five areas shown in Table
I we can conclude that in the United States the per
capita energy consumption is 5.8 times the world
average and 20.5 times the consumption in the Third
World, where most of the petroleum and natural gas
is found. The Third World is in an enviable position
holding 77% of the world's proven crude oil reserves
and consuming very little of it. In fact 72% of the
world reserves are in the hands of OPEC countries,
making this a very powerful coalition.*
At the same time we should realize that the bulk
of the petroleum business is in the hands of seven
multinational petroleum operators (Exxon, Mobil,

---

*OPEC (Organization of Petroleum Exporting Countries):
Indonesia, Iran, Iraq, Kuwait, Saudi Arabia, Abu Dhabi,
Quatar, Libya, Algeria, Nigeria, and Venezuela.

Table I

Per Capita Energy Consumption

Country	Per Capita Energy Consumption in gpd (Gallons of Crude Oil Equivalent/Day)
United States	6.5
Japan	2.5
Western Europe	2.5
Communist Block	1.0
Third World	0.3
World Average	1.1

Texaco, Gulf, Standard Oil of California, Shell-Dutch, and British Petroleum). Thus, reserves, production, processing, and marketing of petroleum products are in the hands of a few nations and even fewer major multinational companies.

The other sources of energy that we have are presently not as important as petroleum and natural gas. Coal is relatively abundant in the U.S.; however, much of it is rather high in sulfur. Research on coal gasification is still at an early stage and much more time and money must be spent investigating the process before it becomes commercial.

Nuclear energy development has been beset with problems involving questions of radiation hazards, safety, and waste disposal to which we do not yet have answers.

Other energy sources such as solar energy and geothermal energy have received only minimal attention from government and industry and little money has been spent on their development.

ENERGY SUPPLY AND CONSUMPTION

The U.S. today has the highest per-capita consumption of petroleum fuels in the world (Table I). Energy consumption in the U.S. has more than doubled in the last 20 years and should double again in the next 20. At the same time, exploration for oil fields, development of them, and refinery building

have not kept pace with this growth. These activities
have been stalled by the environmentalists on one
hand and the higher profits from sale of petroleum
products abroad on the other.

During the next decade most of the growth in the
energy requirements of the U.S. must come from the
import of foreign oil. Natural gas reserves are
declining and imported gas will also assume a growing
share of the market. Coal and nuclear power are not
likely to supply a much greater percentage of our
energy needs during the next decade unless the many
environmental problems are solved.

All this brings us to the problem of self-
sufficiency. Each day we are moving more and more
toward a rather high dependence on petroleum imports.
The situation is certainly not as bad as it is for
Western Europe, which as a block is only 3% self-
sufficient in its petroleum supplies.

Our dependence on oil imports is also affected
by the rapid change from coal burning to oil that
has taken place in Europe and the 2,500% increase
in oil consumption of Japan during the last 20 years.
In fact, although we are presently the prime consumer
of petroleum in the world, our rate of growth in
consumption has been slow compared to other areas
of the world (Figure 1).

If we look at the five major American petroleum
companies (Exxon, Mobil, Texaco, Gulf, and Standard
Oil of California), we find that 10 years ago they
refined 53% of their crude in the U.S. Today, this
figure is down to 38%. This trend must be changed
if we are to have adequate supplies of energy at
home. Accomplishing this will require a large
capital investment, perhaps as high as $200 billion,
and must include construction of additional refineries
as well as expansion of present pipeline capacity.
Because of the large lead times required for this
work, planning and construction of needed facilities
must begin as soon as possible. The American government must oversee the steps taken so that the nation
is guaranteed the supplies of energy that it will
require in the years to come.

A number of measures must be taken to assure
that the energy crisis will not get out of bounds.
To better appraise the magnitude of these measures
I will divide them into short-term and long-term
goals.

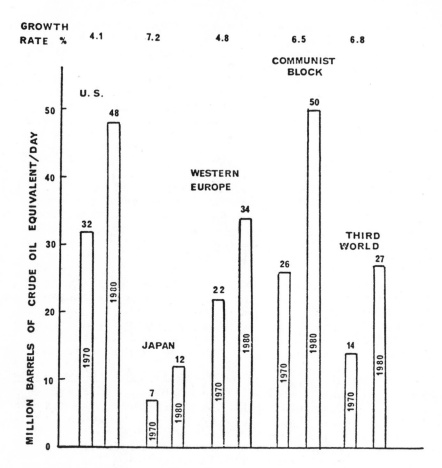

Figure 1.   *World consumption of energy.*

SHORT-TERM SOLUTIONS

The Arab-Israeli war of 1973 disrupted oil flow
in the world and led to the institution of energy-
saving measures in most of the Western nations.
These measures include:

1. increased emphasis on the use of smaller
   automobiles
2. increased aircraft load factors
3. increased industrial efficiency in energy use
4. improved efficiency in power production

  5. improved use of energy in the residential
     and commercial sectors
  6. rationing.

These measures must carry us over until such
times as new energy sources become available and/or
the flow of Arab oil to the Western world is again
normalized.

LONG-TERM SOLUTIONS

A comprehensive national energy policy must be
formulated. This policy should be subject to continuing review and adjustment and should cover the
following categories:

  1. Set up a comprehensive cabinet-level agency
     dealing with all forms of energy.
  2. Change the import quota system that restrained
     the importation of petroleum crude while
     allowing importation of refined products. This
     step has already been taken by the President
     and has served to spur refinery construction
     in the U.S. during the past few months.
  3. Systematize refinery air and water pollution
     guidelines and establish land utilization
     policies to ease the siting of refineries,
     power plants, and petrochemical operations.
  4. Stimulate increases in domestic oil and gas
     production by not regulating gas prices or
     regulating them with realistic price guidelines.
  5. Stimulate increases in domestic oil and gas
     production by increasing the size of offshore
     leases as well as the frequency of lease sales.
  6. Change Internal Revenue guidelines to favor
     exploration in the U.S. rather than favoring
     extensive exploration and production abroad
     by allowing royalties paid to foreign govern-
     ments to be written off against federal tax
     owed the U.S. government.
  7. Foster research on improved coal mining and
     new reclamation techniques.
  8. Sponsor, fund, and encourage coal gasification
     projects to a degree that would lead to pos-
     sible commercialization of the process within
     the present decade.
  9. Foster research on alternative energy sources
     (solar, nuclear, tidal, and geothermal energy).

10. Provide regulations and/or incentives to allow the use of coal as a petroleum substitute in certain industrial operations.
11. Accelerate and systematize the environmental review process of proposed energy projects.
12. Foster research on the environmental problems involved in the production of nuclear energy
13. Aid in the development of commercial stack gas scrubbing.
14. Analyze and implement fuel conservation measures whenever such a need arises.

The objective of such a national energy policy would be to optimize the utilization of our present petroleum resources until alternative energy sources become commercially available in the needed scale.

MINERALS SUPPLY

The metal availability situation is also critical in the U.S.   In 1971 total U.S. metal production amounted to about six billion dollars, and imports were in excess of one billion dollars. In the case of nonmetals, the U.S. has a relatively good position since imports have equaled exports in recent years. The U.S. imports asbestos, fluorspar, potash, sulfur, gypsum, and industrial diamonds while it exports crude fertilizers, natural abrasives and some industrial minerals.

On a global scale one can note that the current value of the world's mineral production is approximately 166 billion dollars; this figure is expected to rise to 520 billion dollars by the year 2000. While the world population has grown 2.36 times since the beginning of this century, the global output of all mineral commodities has grown 12.48 times; therefore the per-capita increase in mineral consumption has been 5.25 times.

WATER

Water has acquired added importance in the last few years due to increased cost and increased demand coupled with increased pollution of our waterways. Schmidt in the last paper of this section discusses the trends in water use projected for the next few years and the potential for water reuse.

2. THE WORLD ENERGY SUPPLY SITUATION

Frank A. Mau
U.S. Department of State
Washington, D.C.

ABSTRACT

Oil, our key fuel resource, has been the subject of
controversy in recent months. In this paper I will discuss the problem of meeting U.S. and world energy demands
over the next 10-15 years. Economic, financial, technological, logistical, national security, and foreign relations implications will be explored. U.S. government
policies dealing with the energy problem will also be
discussed.

The energy situation is presently complicating national
and international efforts to resolve developing ecological
problems. We need to begin using oil and gas in modera-
tion. When it becomes too expensive or too scarce, we
will then shift to some other form of energy.

The emergence of the U.S. as a major competitor for
available world oil is a cause for growing concern in
Europe and Japan, which are almost totally dependent on
oil from the Middle East and Africa. Most experts agree
that to meet projected world requirements for oil, production in OPEC countries must at least double in the next
seven years to about 50 million barrels per day. The
prospect of this happening is not good.

Increases in world oil prices are expected, with these
increases being felt most severely by the less developed
world. A major concern of the U.S. government is to find
a balance between low prices and possible shortages and
higher prices and, hopefully, adequate energy supplies.

The demand for energy in the world has grown
and continues to grow at rates far greater than
those foreseen even in the recent past. This growth
is a matter of great concern to the United States
Department of State, and indeed to the Administration
as a whole. It involves a wide range of national
security, economic, and ecological factors, all of
which must be taken into account in developing a
rational national energy policy. It is clear that
the energy situation is going to complicate considerably national and international efforts to
resolve developing ecological problems.

On April 18, 1973, the President sent an energy
message to Congress. We are all aware of the reasons
for the President's Energy Message. Stated simply,
our energy demands have grown so rapidly that they
now outstrip our available supplies. If our present
rate of growth in demand is maintained, U.S. and
world energy needs will almost double by 1985. In
oil, the current rapid growth in U.S. consumption
indicates that 1980 imports could exceed 12 million
b/d (barrels per day), even assuming that Alaskan
production will contribute two million barrels per
day (Table II). Another 4 million b/d of imported

Table II

U.S. Petroleum Supply and Demand (million barrels per day)

	1970	1980
DEMAND:		
Petroleum for conventional uses	14.8	22-25
Petroleum for gas substitution	--	4
Total	14.8	26-29
U.S. PRODUCTION:		
Lower 48 states	11.3	10
Alaska North Slope	--	2
Total	11.3	12
IMPORTS:		
Western Hemisphere	2.5	4
Eastern Hemisphere	1.0	10-13
Total	3.5	14-17

oil are needed by 1980—increasing to 7 million b/d
by 1985—to substitute for gas that may not be
available in the amounts required if current growth
continues and additional reserves are not found.

THE ENERGY CHALLENGE

The President has noted that we are facing a
vitally important energy challenge. If present
trends continue unchecked, we will face a genuine
energy crisis. Some, including myself, believe that
the crisis is already here. The President has outlined a program that, in his judgment, can lead to
an orderly transition from an economy based primarily
on oil and gas produced in the conventional manner
to an economy supported by a broad range of energy
forms produced at reasonable costs and used in
reasonable ways. He has stated that occasional
shortages of energy will occur and that the transition will not be easy. The program he has outlined
requires action now. It also requires success in a
large number of endeavors where he has indicated
action is required—both domestically and inter-
nationally. Finally, it requires that over the next
one or two decades adequate supplies of oil and gas
be available to meet the needs of the U.S. and the
world while we move towards alternative forms of
energy.
    Even though the U.S. represents less than 6% of
the world's population we consume one-third of the
world's energy. World reserves of conventional oil
are believed adequate to meet world demand for oil
until alternative energy sources are developed. The
bulk of presently known reserves are in the hands of
OPEC nations, particularly those in the Persian Gulf.
Within these parameters, a solution to the energy
problem seems simple. We use oil and gas in modera-
tion, including imported oil and gas, until it becomes
too expensive or too scarce and then we shift to some
other form of energy. That is precisely what we will
do. Since this is the case, why are we in Washington
so concerned?
    Does government really have to enter into the
problem? Some say no. Remove all government controls
(which got us into the problem) and the Law of Supply
and Demand will solve any further problems. Others
say that the government must play a major role,
instituting policies designed to insure a continuing

and adequate supply of energy to the consumer at moderate prices without damage to the environment. Unfortunately, the problem is not that simple. The government already is involved in the energy business and the majority of the American people appear to expect it to continue to play a role. But, do they want or expect the government to become the principal supplier of their energy needs? A free world market in energy does not exist, with a growing portion of the world's oil and gas being marketed by sovereign states. Can the private sector adequately defend the interests of consumers in dealing with oil-producing governments? Can moderate prices be maintained? Will enough energy be available? Can our environmental concerns be entirely met? Will alternatives to oil and gas be available in time? Let us attempt to look at these questions in some detail.

Will alternatives to oil and gas be available in time? In the medium term (*i.e.*, the next 10-15 years) most government officials see no possibility that significant reductions can be made in our oil requirements through alternative energy sources or reduced consumption. A major effort must be made, beginning now, to reduce consumption and to develop alternatives to imported oil. However, like it or not, we and the rest of the world are stuck with an increasing requirement for imported oil over at least the next 10 years, and probably for an even longer period. Nuclear power has fallen far behind schedule. Drastic policies will have to be adopted if atomic power, coal, oil shale, geothermal, solar or other alternative forms of energy are to make a significant contribution—or even hold their present share—of U.S. energy demand. The President has announced a program to move towards alternatives to oil and gas. Will it be enough?

Can our environmental concerns be met entirely? Regardless of whether our oil or gas is imported or produced domestically, new refineries and pipelines must be constructed. Superports will also be required, and sites must be found for hundreds of new atomic power plants. If we are to avoid a substantial dependence on imported oil, drilling must be conducted on a massive scale in Alaska, off both coasts and in the interior of America. Some of the proposed remedies one hears to our environmental problems would involve major changes in our style of living and massive disruptions of our economy. Are we ready for these changes? If not and if the energy problem is to be met, many compromises in the environmental area must be reached. These will require additional hard decisions and much courage.

AVAILABILITY OF ENERGY

Will enough energy be available? I already have suggested that, in the short to medium term, alternatives to oil and gas will not be available. Also world reserves of oil appear ample to meet world requirements until other forms of energy "come on stream." The problem is that it may not be made available in the amounts required, the price may be high and the balance of payments and security risks involved may be considerable.

The emergence of the United States as a major competitor for available world oil is a cause for growing concern on the part of Europe, Japan, and other consumers. Japan imports almost 90% of the oil it uses from the Middle East and Africa. Almost 80% of Europe's oil comes from the same sources. This level of dependence will not decrease substantially over the next ten years. Meanwhile, their own and U.S. supply requirements will increase enormously. Dependence on Middle East oil reserves is unlikely to diminish in this period even though exploration and exploitation are carried out in many parts of the world. Two million b/d from Alaska will only compensate for our declining production in the Lower-48 states. A possible 4 million b/d from the North Sea will represent less than one-sixth of Europe's requirements by 1980.

The United States is the world's largest producer of oil. In March 1972, we essentially went to the maximum efficient rate of production in our domestic fields. Unless large new fields are found, or expensive secondary recovery operations are undertaken on a massive scale, future production will decline. Both must be attempted, but most experts meanwhile agree that to meet projected world requirements for oil, production in OPEC countries must at least double in the next seven years to about 50 million b/d. Is this likely? Production already has been curtailed in Kuwait, Libya, and Venezuela. The Shah of Iran has suggested that production in Iran will be held, for reasons of conservation, to eight million b/d. Production in Iran is currently about six million b/d. The Canadian National Energy Board, in

a recent public document, suggested that Canada may
have no conventional oil available for export to the
United States by the end of this decade if future
Canadian needs are to be adequately met on the basis
of presently proven Canadian oil reserves.

Many in the oil industry believe that only Saudi
Arabia and, possibly, Iraq have sufficiently known
reserves of oil to meet expanded world needs. Some
projections suggest that Saudi Arabia alone must
produce at least 20 million and, perhaps, as much as
35 million b/d by 1980 or shortly thereafter if world
requirements are to be met. The United States Department of State has serious doubts that Saudi
Arabia, for entirely valid reasons unrelated to
politics, will permit production at the latter rate—
almost 10 times current production in Texas.

FOREIGN POLICY AND OIL

Other major consuming nations are concerned about
the implications of U.S. foreign policy, particularly
our Middle Eastern policies, on the future supply of
oil. During his visit to Washington early in 1973,
the Saudi Minister for Petroleum, Zaki Yamani, made
reference to this subject. It is the firmest intention of the United States government to work toward
resolution of the Arab-Israeli problem, but even if
it is resolved to everyone's satisfaction (an unlikely event), we still have an energy problem. Our
relations traditionally have been and continue to be
friendly with the major oil producers of the Middle
East. They must remain so. The more responsible
OPEC governments, including those in the Middle East,
have recognized that producers and consumers alike
must work toward some kind of an accommodation to
assure adequate supply. This is also our goal.
However, we must recognize that imports from the
Eastern Hemisphere as a whole, particularly from the
Middle East, are potentially unstable. This must be
considered in our planning.

An important question is what is the future of
the relationship of the United States with the major
oil-producing and -consuming countries? A number of
initiatives are under consideration in Washington
and other capitals. It is not misleading to say
that we favor cooperation among producers and consumers to meet the mutual problems created by
shrinking world oil supplies. However, it is one
thing to advocate cooperation and another to achieve

it in a way that produces positive results rather than deepened confrontation. Real problems and obstacles of interpretation and individual national interests and need exist. The formulation and execution of energy policies in the international arena will be one of the most important and demanding tasks of our own and other governments over the next decade.

Can reasonable energy prices be maintained? Let's look at this first in the international context. Shortages of energy materials have affected Europe and Japan to a considerably greater extent than the United States since they have fewer alternative energy sources. Rapidly increasing energy costs will be of crucial interest to all users of energy, but the effects will be most severely felt by the less developed world. India was not able to absorb the 30 cents per barrel increases in oil costs resulting from the Teheran agreements between OPEC and the oil companies and had to cut back its consumption of imported oil proportionately. Most observers expect increasing prices over the next decade under almost any circumstances. If, however, there is a physical shortage of oil in the world and the main industrialized countries are competing for the oil offered for sale, the price can rise spectacularly.

OIL PRICE INCREASES

Increases in world oil prices are of direct concern to the American consumer as we become more dependent on overseas supplies. Both increased prices and increased imports represent a nightmare for those in Washington concerned with balance of payment matters. Events of 1973 indicate the vulnerability of the world monetary system to shifts of relatively few world currencies. The potential for mischief, whether deliberate or otherwise, is enormous as massive amounts of money are transferred to the oil-producing countries.

The inevitable rise in the cost of energy, whether produced domestically or abroad, is a cause for concern to us all. It has a direct bearing on our ability to compete abroad and on the fight against inflation at home. A major preoccupation in Washington these days is to find a balance between low prices and possible shortages and higher prices and adequate supplies of energy.

Can the private sector adequately defend the
interests of consumers in dealing with the oil-
producing governments?   This is an awkward question
to ask in a free-enterprise society, but it must be
faced.   There is a growing body of opinion in Europe
and Japan that consumer governments must have a voice
in future negotiations with the producing countries,
either directly or through the oil companies, by
advising on options and on the extent of commitments
the companies should or should not make.   There have
been suggestions that the consuming countries,
through some formal or informal mechanism, should
attempt to enter into direct negotiations with OPEC
nations to insure the future availability of oil in
the quantities required and at reasonable prices.
In any case, it is clear that the world oil industry
is in a state of evolution with the outcome being
difficult to predict.   Early in the next decade, if
not sooner, over one-half the oil moving in world
trade will be controlled directly by the oil-producing
governments.   Among the hardest questions of our time
to answer will be the degree to which government
should or must be permitted to insert itself into
activities formerly reserved entirely to the private
sector.

IS GOVERNMENT CONTROL NECESSARY?

This brings us back to our starting point.   Do
the American people want or expect the government to
take principal responsibility for supplying our
energy needs?   Many apparently do, but as everyone
knows this is not really possible.   The President
has asked the Congress and the American people to
move with him toward resolution of our energy problems.   Some of his recommendations have and will
give cause to controversy.   His leadership is essential and government must show the direction.   But
only we, Americans acting together, can really
effectively meet the challenges of the energy crisis.
The production and consumption of energy are
certainly not the only causes of pollution, but they
are major sources, with important effects arising
from almost every incident of these functions.
Strip mining of coal and the utilization of platforms
at sea for oil drilling are increasingly necessary
in connection with the extraction of adequate quantities of fossil fuels.   The danger of leakage and
spills, as well as unsightly pipelines and power

lines, are some of the problems in the transportation
of fossil fuels and energy. Increasing dangers occur
as economic factors result in the use of larger
vessels and pipelines over dangerous terrain. The
combustion of fossil fuels results in varying degrees
of air pollution, for which adequate controls are
not yet economically available. The advent of
nuclear power will result in thermal effects on
neighboring waters and radiation, the effects of
which have not yet been fully evaluated. The disposal of radioactive wastes from nuclear energy
will also present important problems, as will the
disposal of wastes resulting from the production of
synthetic gas and oil from coal.

These environmental problems of the extractive
industries will not go away. They cannot be
answered by "zero growth." They and their associated problems of energy supply and costs and
conservation are real and honest problems for the
U.S. and the world. We must seek now real and honest
answers if the U.S. and the world, as we know them,
are to move forward—or stay even.

3. WORLD MINERAL OUTLOOK

Dean Beckerdite
Minerals, Resources,
and Materials Department
Standard Research Institute
Menlo Park, California

ABSTRACT

   The mineral industry may, one day soon, be truly
global. This geographic expansion will be accompanied
by economic, social and political implications for which
there is no precedent. Even in familiar geographic
areas, the industry is being forced to cope with social
and political issues that are increasingly outside its
sphere of influence. To be successful during and after
this period of unsettled conditions, mineral companies
must identify those specific changes that may affect
their economic future and then act accordingly to
minimize risk and optimize profit. Long range planning
efforts have never been more demanding or more necessary.

   The evidence of dynamic change occurring in the
natural resource industries is overwhelming. In
world minerals, two changes are the most outstanding.
First, the geographic expansion of the minerals industry has reached proportions where the industry
has become truly global, and for this there is no
precedent; second, even in familiar geographic areas,
the industry is forced to cope increasingly with
social and political problems, problems that are
outside the industry's sphere of influence. As a
result, the successful mineral producers' planning
efforts are becoming more and more complex. Fifteen
years ago, long range planners were primarily concerned with demand forecasting. Even ten years ago

19

it was common for planners to assume that supply
would be plentiful and would increase with demand.
Five years ago, the word *environment* was used to
describe business conditions. Three years ago, we
assumed the dollar would still be a dollar—relative
to other currencies. As recently as one year ago,
it was often assumed that fuel and/or power would be
available at "reasonable prices." There have been
many other changes, and resource management has
found that they all have one common element—increasing complexity. Resource management decisions
have never been more difficult.

Even without these new complications, the
minerals industry was already complex enough. It
produces more than 70 commercial metallic and non-
metallic minerals. Frequently, a single mineral is
the basis for an entire industry that varies considerably in terms of size, growth, structure,
geographic emphasis, profitability, and technical
requirements. Furthermore, for different metals,
supply, demand, and price outlook can and do vary
considerably.

Recognizing the conglomerate nature of the
minerals industry, let us now review some significant
annual growth rates in production over the past 12
years. Metal production of platinum, aluminum ingot,
vanadium, molybdenum, nickel, and titanium has increased at a rate of more than 5% per year. Metal
outputs of zinc, tungsten, iron ore, chromium, lead,
manganese, copper, tin, silver, mercury, and gold
have increased slower than 5%. Actually we can note
that (demand or consumption) the supply of gold,
silver, or copper has not increased.

Among the nonmetals, production of fluorspar,
potash, phosphate, talc, cement, and salt has increased faster than 5% while outputs of sulfur,
asbestos, feldspar, barite, and gypsum have increased
slower than 5%.

The nonpetroleum fuels have not done very well.
For instance, coal production has increased slightly
while uranium production has actually declined.
This, I think, demonstrates the national complacency
that preceded the current fuel shortage. Virtually
all the energy demand increases of the previous
twenty years have been satisfied with oil and gas,
a pattern that cannot continue forever.

By volume, only six world minerals exceed 50
million metric tons of product annually. Coal, at
more than 3 billion tons, is more than four times
larger, by volume, than the nearest competitor—iron
ore. A look at price indicates that growth in

consumption has not been a dominant factor.　Although most mineral prices have increased, price has actually declined in some specific cases (Table III).

Table III

Mineral Commodities that have Declined in Price

Aluminum - The aluminum industry is market oriented, and dependability of supply is essential when attempting to increase market share.　Overcapacity is essential to assure supply dependability which, in turn, causes downward price pressure.

Asbestos - For medical reasons, asbestos usage is declining. Substitute materials are making market inroads, and overcapacity exists.

Barite - Drilling mud demand has declined with a reduction in oil drilling and overcapacity exists.

Manganese - New discoveries have created a world surplus, and price/demand elasticity is virtually nonexistent.

Phosphate - Phosphate production capacity increased too fast— a situation totally unrelated to the resource base.　The result is that this very desirable product is being exploited too fast, because of an industry planning error.

Potash - One of the most successful exploration efforts of our time created a surplus.　Canadian reserves are enormous.

Salt - This commodity is too plentiful and very compatible with economy of scale production technology.

Sulfur - Overcapacity caused by air clean-up activities, and other by-product production has created a world surplus that is destined to continue.　The worst is yet to come for suppliers.

Talc - Usage of talc has declined and the result is overcapacity.

Titanium - The SST was not built and therefore projected demands did not materialize leading to overcapacity.

Uranium - The ambitious development programs of the past created a short-term surplus—by 1995 we may wish this matter had been handled differently.

It is important to note that resource base and planning have long-term predictable impact on resource price. At least the direction of price change over the long run can be forecast.

For all the sophistication in mineral management, the discounted cash flow philosophy wins out in the end. *The major factor affecting current price is short-term availability.* The facts are that, for the most part, basic resources are so important that price only becomes a factor after availability is established. Price volume sensitivity is nil for most resource commodities. For example, 8 lbs of fluorspar are needed per ton of steel. No more and no less can do the job. Similarly, exactly 0.6 lbs/ton of vanadium are essential in steel making.

In general it is apparent that prices for mineral resources are usually established without any thought of replacement cost. The resource base *has not* been given sufficient consideration by resource managers. In the past, it has been too easy to assume that resources will be found on an "as needed" basis, an attitude that is getting us in trouble. Exploration takes time, money, and planning.

Management must explore and realize the special significance of time. It takes at least seven years for most companies today to materially improve their resource reserve positions. Seven years is about the minimum required time for management to improve its resource base and make products available for sale and use.

We simply *must* recognize that our natural resources are not unlimited, and that low or reasonable price is not necessarily an indication of shortage. The resource base, while different for each resource, is established by exploration and technology, not by marketing or financial tricks. Although exploration may not seem attractive via discounted cash flow analysis, the alternative is to deplete existing mines and go out of business, which is not a very satisfactory alternative.

The basis for a profitable mining venture is a good mine. New good mines will always be developed at the expense of marginal mines. Before a mining firm can tell whether or not it has a good mine, it must know a lot about other mines, wherever they occur. The resource producer is engaged in an international business, and he can not afford to ignore worldwide developments. He is affected by them every day.

CONCLUSIONS

Throughout this brief discussion of world minerals there have been two rather basic thoughts. In conclusion, I would like to emphasize them again:

    1. Failure to consider global developments in planning can be disastrous for a resource producer.
    2. The incentive to find and develop new mines has never been greater. Massive exploration efforts are necessary, now, if we are to avoid a future of constant critical resource shortages.

At Stanford Research, we correctly predicted the energy crisis well in advance, simply because new reserve discoveries were not keeping pace with consumption and because trends in industry concentration were threatening the traditional industry structure.

We are now warning of future crises in a few of the world mineral commodity areas—especially where balance of payment problems and trends toward nationalization will affect traditional source relationships.

Each of us has a responsibility to do whatever we can to assure that the U.S. energy crisis is not the first indication of a *total resource crisis*. For whatever reason, resource planning can not be allowed to fail us again.

4. THE IMPACT OF ENVIRONMENTAL LEGISLATION ON WATER RESOURCE UTILIZATION

Ronald G. Schmidt
Wright State University
Dayton, Ohio

ABSTRACT

Water usage has continued to increase sharply in recent years. Pressure to reverse the downward trend in water quality has resulted in the enactment of both federal and state legislation. Review of federal water legislation reveals a consistent trend toward increased federal control. More stringent standards have been imposed simultaneously. The Federal Water Pollution Control Act Amendments of 1971 and 1972 have reached a culmination in pollution control of surface waters. They call for a policy of cessation of all discharges by 1985. "Zero Discharge" implies total water reuse, with such reuse reducing water withdrawals drastically and improving surface water quality to pretechnology levels. Whether or not Zero Discharge can be attained depends on the development of reuse methods and equipment. There is considerable research in these areas at present, but there are a number of priority areas that should receive greater attention.

WATER USE PERSPECTIVE

Upwardly spiraling technology, continuous growth in standard of living, and continually increasing population together have led to exponential growth in our use of water. In 1960 (see Figure 2) water withdrawals from both surface and ground water reservoirs amounted to approximately 250 billion gallons/ day. Much of this water can be reused and indeed

Agriculture	139 billion gallons/day
Power (Steam Electric)	66
Manufacturing	23
Municipal	20
Mining	3
	251 billion gallons/day

Figure 2.   U.S. water withdrawals 1960 (Data from Wollman and Bonem[1])

is at least once if not multiple times simply by return to the source. Approximately half, that is, the principal portion of the 139 billion gallons/day utilized by agriculture and certain factions of both the steam-electric power and manufacturing portions, are consumptive in the sense that they are not returned to the source but are evaporated, transpired or infiltrated into some other portion of the hydrologic cycle. In 1971, Wollman and Bonem[1] published the results of a comprehensive study of U.S. water supply prospects in relationship to projected demands. Because both the available supply and the demand have considerable variables related to time, quality, economic growth, population and other factors, the measurement of either supply or demand even today is a complex undertaking. To make such projections for the future requires both consummate skill and divine insight. Nevertheless the study produced by Wollman and Bonem goes a long way toward developing a rational framework within which to make long-range water management decisions. Comprehensive variables involved are dealt with in the study but boil themselves down to a basic model that calls for an instream dissolved oxygen standard of 4 milligrams of oxygen/liter of water and a 98% availability of flow (or a 2% chance of deficiency). Under these conditions projections of growth and of water withdrawals are presented in Figure 3. The high estimates presume different rates of economic population growth than do the medium and low estimates. The need to more than double our use of water in the next 50 years is coupled with a parallel need to maintain or improve significantly the quality of that water.

Projections of national need are meaningless when the very high costs of interbasin transfer of water make adjustment in regional deficits generally an unlikely way of equalizing regional demand with

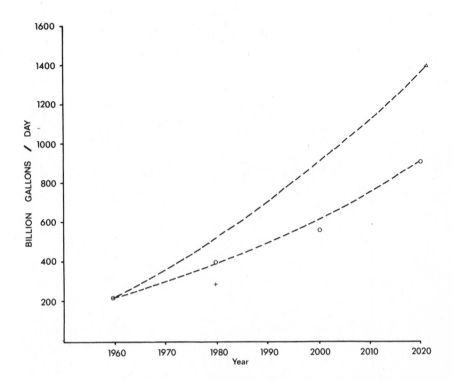

Figure 3. U.S. projected growth in water withdrawals (data from Wollman and Bonem[1])

regional supply. Despite the recent successful completion of the California Water Plan, interbasin transfer is not likely to become a general practice. Table IV shows the maximum regulated flow and projections of demand for 1980, 2000, and 2020 by regional water basin. It is clear from these estimates that the upper Rio Grande-Pecos, the Colorado, the Great Basin, and the South Pacific Basin will all be water-deficient by 1980. Strong demand pressure will have to result in major economic investment and decreased consumptive use, increased storage and reduced water contamination, thus making more water supplies available within the basin or alternately to undertake high cost interbasin transfer. Strong pressure from population or economic growth will result in water demand close to or at the high forecast levels. By the year 2020 nearly half of the basins will be water-deficient, and the

Table IV

Water Resource Regions[1]

	Max. Regulated Flow	Medium			High
		1980	2000	2020	2020
		(billion gallons/day)			
New England	60,895	3,177	4,522	6,474	9,935
Delaware and Hudson	28,629	6,486	9,785	14,627	25,907
Chesapeake Bay	46,657	6,025	10,410	17,767	39,329
Ohio	99,457	4,154	6,748	11,055	23,041
Eastern Great Lakes	33,278	4,800	7,995	13,482	30,471
Western Great Lakes	30,283	10,639	17,502	30,641	71,965
Upper Mississippi	46,125	3,350	5,321	8,275	16,133
Lower Missouri	16,211	957	1,657	2,896	5,703
Southeast	186,030	25,451	48,176	87,941	186,781
Cumberland	14,647	1,810	4,280	9,088	23,529
Tennessee	40,389	3,019	5,742	10,381	24,493
Lower Mississippi	35,207	3,130	5,311	8,536	16,732
Lower Arkansas-White-Red	57,661	3,099	4,463	6,064	10,114
Upper Missouri	25,600	15,912	18,179	24,084	38,553
Upper Arkansas-White-Red	7,053	6,730	7,486	8,969	14,550
Western Gulf	25,900	17,235	26,737	44,441	98,408
Upper Rio Grande-Pecos	3,000	5,507	6,529	8,921	12,901
Colorado	11,400	16,950	25,204	42,643	65,373
Great Basin	6,934	6,251	7,011	10,046	18,038
South Pacific	815	8,135	12,278	18,055	26,098
Central Pacific	45,478	26,834	30,309	37,267	54,872
Pacific Northwest	134,570	25,068	36,886	58,005	96,342
United States	956,219	204,719	302,541	470,658	909,268

total water withdrawal to meet the demand will be
very close to the maximum regulated flow figure of
956 billion gallons/day.
      What does all this mean in terms of the indi-
vidual, industrial, municipal or agricultural user
of water? Since water supply is so closely tied to
standards of water quality and the maximum regulated
flow estimates made are in terms of the dissolved
oxygen and flow reliability constraints mentioned
previously, flow regulation (that is, storage) will
be greatly affected. Therefore water supply will be
greatly affected by improvements in water quality
due to improved waste treatment facilities and/or
water reuse.

WATER QUALITY

      The principal problem of increasing water supply
becomes a question of improving water quality. And
it is for this reason as well as consideration of
public interest, recreation and aesthetics that
there have been local, regional and federal, public
and governmental pressures to improve water quality.
      History of governmental concern for water quality
goes back to 1886 when federal legislation was passed
to protect New York from the dumping of refuse (see
Table V). This initial, tentative effort was followed by the Federal Refuse Act of 1899, which was
designed generally to protect rivers and harbors
against prohibitive discharge of waste materials
into the water. Although originally designed to
protect shipping from the wholesale distribution of
floating debris, the Act has most recently been
utilized to prevent or control the discharge of
chemical and sanitary wastes into navigable waterways.
      In 1912 the Public Health Services Act was passed
to investigate the effect of various pollution constituents in lakes and streams on public health. In
1924 the proliferation of waste associated with oil
production stimulated the passage of the Oil Pollution Control Act, again to protect rivers, harbors,
and docks from waste discharge related to oil production, distribution and refining. Being preoccupied
with economic recovery from the Great Depression and
the Second World War, federal legislators were inactive in this area for 24 years until 1948, when the
first Water Pollution Control Act was passed to work
cooperatively with state governments in the investigation and regulation of various aspects of pollution

Table V

Historical Summary of Federal Water Legislation

Year	Law	Affected	Remarks
1886	---	New York Harbor	Prohibited dumping of refuse in New York Harbor
1899	Refuse Act	Rivers and harbors	Prohibited discharge of waste materials into navigable waters
1912	Public Health Service Act	Lakes and streams	Investigate effect of pollution on public health
1924	Oil Pollution Control Act	Rivers, harbors, docks	To control oil waste discharge from production
1948	Water Pollution Control Act	Interstate waters	Cooperative with states
1956	Water Pollution Control Act - Amended	Interstate waters	States have primary responsibility authorized research, facilities, grants
1961	Water Pollution Control Act - Amended	All navigable waters	Strengthened 1956 Act
1965	Water Quality Act	A "national policy" for regulation of water pollution	Created Federal Water Pollution Control Administration in HEW - FWPCA moved to Interior, then later to EPA (1970)
1966	Clean Water Restoration Act	International boundary and inland waters	Together strengthened 1956 Act, increased expenditures, imposed time-table for cleanup
1967	Water Quality Act - Amended	---	States given until June 30, 1967, to adopt water quality standards for interstate waters
1970	Water and Environmental Quality Improvement Act	Maine and coastal areas rivers and harbors	Control spillage, especially oil Supplements 1967 Act
1971 & 1972	Federal Water Pollution Control Act Amendments	Navigable Waters	Zero discharge by 1985-1977 & 1983 interim deadlines. Public participation policy. Applies to point-source.

with respect to interstate waters. Amendments to
this Act in 1956, 1961, and 1965 strengthened and
extended support responsibility to all navigable
waters and made water quality a national concern.
In addition major federal administrative machinery
was set up in the form of the Federal Water Pollution
Control Administration. More recent amendments in
1966, 1967, and 1970 have further extended the areas
of control to all interstate waters, navigable or
otherwise, including inland and international
boundary streams, lakes, and shores. The effect
through this sequence of legislation has been to
strengthen step by step the federal role in pollution
control and, although state sovereignty is recognized,
actual authority has been decreased. In 1971 and
particularly in 1972, the Federal Water Pollution
Control Act Amendments have reached a culmination
in pollution control of surface waters.

FEDERAL WATER POLLUTION CONTROL
ACT AMENDMENTS IN 1972

On October 8, 1972, Public Law 92-500 was enacted,
establishing new national goals for the elimination
of all pollution from all discharges into the waterways and creating broad avenues for public participation in the federal and state programs designed to
bring about these goals. The Act states:

> The objective of this Act is to restore and
> maintain the chemical, physical and biological
> integrity of the Nation's waters. In order to
> achieve this objective it is hereby decreed
> that consistent with the provision of this
> Act . . . it is the national goal that the
> discharge of pollutants into the navigable
> waters be eliminated by 1985.

The establishment of a national goal of "Zero
Discharge" signals a major departure in water pollution control objectives for both state and federal
agencies. Past legislative programs and policies
have emphasized the assimilative capacity of streams
and their ability to dilute pollution. The new goal
is to maintain the natural, biological and ecological
integrity of standing and flowing water and to decrease and ultimately eliminate its use as a receptacle
and transporting medium for wastes. Additionally it
provides that "it is a national policy that the

discharge of toxic materials in toxic amounts be
prohibited." Implementation of the Act involves the
achievement of certain standards by a series of
three deadlines. It additionally provides for a
number of mechanisms whereby these standards will
be achieved.

Under the provisions of the 1972 Act, effluent
dischargers were required to have made formal application by early 1973 for a permit to continue discharging effluent. A National Pollutant Discharge
Elimination System (NPDES) is provided in the Act to
regulate discharge of pollutants by all point source
dischargers. Nonpoint source discharges such as
agricultural runoff, construction contaminants and
drainage from mining regions are not regulated under
NPDES.

Although the 1972 Act does not revoke the 1899
Refuse Act, it charges the Federal Water Pollution
Control Administration (FWPCA) with the responsibility for administering the permit system under
the control of the federal EPA and the states, thus
effectively replacing the Refuse Act program adminis-
tered by the Corps of Engineers. Permits granted
to industries and public-owned treatment facilities
must observe toxic effluent requirements, new source
performance standards, and requirements based on
water quality standards by three deadline dates:
1977, 1983, and 1985. By each of these dates per-
mittees will be required to improve the quality of
their discharge in accordance with very strict
state-of-the-art capability determinations. The
permits, when issued, will contain effluent limitations, schedules of compliance with respect to
construction or process change schedules in order
to meet deadlines in the phased manner. Self-
monitoring and reports are requirements under the
Act. Violations of any of the provisions, deadlines
or requirements are specifically expressed as violations of the Act. All reports will be publicly
available. The Act is unique in that it provides
specifically for public participation as a national
policy.

> Public participation in the development,
> revision, and enforcement of any regulation,
> standard, effluent limitation, plan or program
> established by the administrator or any state
> under this Act shall be provided for, en-
> couraged, and assisted by the administrator
> and the states. The administrator in coopera-
> tion with the states shall develop and publish

regulations specifying minimum guidelines
for public participation for such processes.

Guidelines for such participation have already
been proposed by EPA. Standards to be achieved by
each of the deadlines (1977, 1983, 1985) are expressed
in terms of the availability of pollution control
technology for the specific effluent character.
Industrial sources of pollution must achieve "best
practicable control technology currently available
by July 1, 1977." By 1983 they must meet the "best
available control technology economically available"
and, in addition, municipal facilities must achieve
"best practicable waste treatment technology." By
1985, the national goal is Zero Discharge. Unlike
previous water pollution control programs that related industrial and municipal waste clean-up
schedules to specific water quality standards and/or
to specific uses for which a stream is classified
and/or the assimilative capacity of that stream, the
new standards require that the best existing technology at the time ("state-of-the-art") must be
applied for water quality improvement even if this
results in stream discharge better than the existing
quality of the stream. It is therefore clear that
it is the technology available and existing at the
time for each industrial category that will determine
the requirements for any individual plant.
The Act also makes clear that individual plants
in the same manufacturing category will be required
to attain the same standards no matter where they
are located, thus preventing plant movement pressures
caused by unequal application of standards from state
to state. All of the deadlines and attending standards are clearly designed to encourage reuse technology and ultimately to prevent discharge into
public waterways. The discharge of toxic pollutants
is entirely prohibited under the new law and a very
strict schedule for elimination was proposed by EPA
by mid-1973. New sources that became operational
by January 1974 were required to meet technological
standards based on "best available control technology
adequately demonstrated," as defined by EPA for each
major industrial category.
Industries that transmit their wastes to municipal
or regional plants for treatment will be required to
meet an additional set of standards called "pretreatment standards." These will prevent such wastes from
damaging or merely passing through the treatment
facility. Wastes like hexavalent chromium that are
toxic to biological treatment bacteria and/or

constituents that are not biodegradable will not be permitted to enter the plant.

The initial issuance of a permit and the 1977, 1983, and 1985 required standard achievement deadlines will involve public hearings, public notices, and sufficient time for input by both publicly and privately interested parties. Under the NPDES provision, states may apply for both interim and permanent authority to operate their own permit programs. Several states, including Ohio, have already done so. In order to receive such authority from EPA on a permanent basis they must submit evidence of enabling legislation and the establishment of sufficient administrative and enforcement machinery to be able to carry out the provisions of the 1972 federal act. If state statutes differ from the 1972 Act, they must be altered by the state to provide equal to or more stringent standards before permanent authority is delegated. Specific guidelines for delegation of such authority to the states have already been published by EPA. Decisions by specific effluent dischargers concerning individual plant operation discharged to a municipal or regional system or any other means of treatment or disposal of wastes must be based upon statewide, basinwide, or areawide water master plans.

Under the Act, every state is required to have a continuous planning process for all navigable waters in the U.S. in order to participate in NPDES. Initial plans for implementation of the planning process provisions have already been requested and when approved will require statewide annual reports to EPA concerning priorities, manpower, and other resource needs. Sections of the Act require areawide planning, not only for sources of point-source pollution but also for nonpoint source pollution as well. Once an approved areawide plan is in effect, grants for construction of waste treatment facilities will be made only for those facilities that are in conformance with the plan.

Enforcement of the provisions of the Act are by two basic means: either federal or state authority, depending upon whether that authority has been delegated, and the exercise of the inherent regulatory provisions, or second, by citizens' suits. Under the Act, citizens are permitted to bring civil suit against any person, industry, or agency alleged to be in violation of the effluent standard limitation or order issued by an administrator of the state or federal EPA or even against the administrator if he has allegedly failed to perform his duties under the Act.

ALTERNATIVES FOR COMPLIANCE
WITH THE 1972 ACT

Review of the 1972 Act makes it clear that significant progress in environmental improvement will
be required on a rather strict time schedule for
everyone involved in waste discharge. Whether or
not Zero Discharge can be achieved by 1985 remains
to be seen, but Zero Discharge has been established
as a national goal and waste management programs
will be directed toward that ultimate goal. It is
apparent that waste treatment measures involving
return to interstate waters will be interim and,
except in rare cases of urgent need, probably not
worth the investment. What are the major alternatives
then? Under what conditions do they apply? What
are their limitations?
In the present framework of wastewater treatment
technology used water can be treated by two basic
means (Figure 4). Biological treatment is that process in which wastes containing principally organic
constituents (such as sanitary wastes), after a
process of comminution and grit removal, are exposed

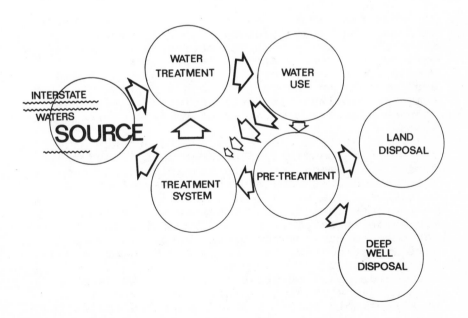

Figure 4. Treatment of used water.

to consumption by several strains of bacteria. Given sufficient oxygen and the proper environmental conditions, these bacteria proliferate and consume waste products. After treatment the effluent is clarified, and in the case of advanced treatment, the main constituents such as nitrates and phosphates are removed by various chemical means. The presence of significant quantities of inorganic chemicals and/or nonbiodegradable organic constituents may require extremely sophisticated equipment, and in some cases, further chemical processing. Removal of trace quantities of heavy metals or other undesirable constituents can be achieved at the present time only through very expensive processes. As provided under the Act these kinds of materials will have to be removed through a pretreatment process.

As an alternate to the long-established biological treatment plant and rapidly growing in utilization is the physical-chemical treatment plant. Essentially it is a chemical refinery operating by a variety of inorganic processes to remove various constituents from the water. Other methods of handling of wastes may be considered if ultimate disposal does not include return to interstate waters. Depending upon the characteristics of the wastewater, various kinds of pretreatment may be employed, and the resulting effluent can be disposed of either by spray irrigation on land surfaces or by disposal through injection in deep wells. In the case of ultimate disposal on land surfaces, considerable pretreatment is necessary to reduce or eliminate suspended solids and possibly to remove toxic constituents. Effluent still containing principal nutrient constituents can then be used for irrigation of forests or recreational areas.

Application to land surfaces to enhance growing crops may be possible if pretreatment includes provision for removal of human and animal pathogens. This method of ultimate disposal has rather wide tolerance for a variety of waste constituents. Its limitations include the necessity for controlling discharge in response to soil and climatic conditions, the need to monitor the migration of fluids, and ultimate of course, the increased consumptive use of the water.

The disposal of pretreated wastes in deep wells in many cases permits a minimum of pretreatment, usually simply the removal of suspended solids, perhaps some chemical adjustment or concentration. This disposal method is really not so much a means of disposing of effluent as of storing it in

underground space, isolated from the present and
anticipated future contact with the ecosphere.
Waste stored in this way may, under certain conditions, be retrieved later if this is desirable.
This disposal method permits maximum tolerance of
quite variable chemical constituents including
extremely toxic materials. In general it can be
accomplished economically even today. The limitations of the method include a requirement for
relatively concentrated (*i.e.*, low discharge) waste
effluent and the need to monitor and carefully control the placement and ultimate migration of the
waste from the point of injection.

The ultimate objective for disposal of effluent
is, of course, reuse. Depending upon the characteristics required for water use, wastewater can be
either pretreated or more fully treated short of the
ultimate treatment necessary for return to interstate
waters and then returned as input to the water use
cycle. It is probable that the greater majority of
water uses could employ renovated water on a recycling basis. Under the pressure of alternate
higher level treatment required for return to
interstate waters and with the additional economic
incentive related to the reclamation of useful or
valuable wastewater constituents, such recycling
and reuse can be an attractive objective. Under
such conditions, a water user will require only that
additional supply necessary to make up shortages in
his recycle system resulting from the consumptive
use that might be involved. Limitations, of course,
include possible concentrations of undesirable
constituents and the treatment necessary to remove
them.

Other alternatives not presently under wide use
and not included in this discussion include sewage
distillation and dehydration. The economic and
technological feasibility of these are insufficiently
clear at this time.

RESEARCH

Recognizing certain technological limitations
inherent in the 1985 Zero Discharge objective,
federal and state governments are undertaking in-
house or are supporting externally significant
research aimed at providing answers quickly. Research target areas for the federal EPA include:

1. *health effects* including epidemiological studies, toxicological, and teratological research
2. *ecological effects* including studies of terrestrial and aquatic ecosystems both biotic and abiotic as impacted by pollutants and toxic substances
3. *transport processes*, meteorological research and pollutant pathways, physical and chemical phenomena, and eutrophication
4. *measurements and instrumentation* including analytical methods development and trace contamination detection instrument development
5. *water supply* including research on drinking water standards, protection of tapwater quality and water treatment
6. *municipal water control technology* including advanced waste treatment, water renovation, combined storm and sewer overflows
7. *industrial water control technology* including industrial water pollution control
8. *nonpoint source water control technology* including mining and agricultural water pollution control, control of oil and hazardous materials spills
9. *monitoring* including measurements standardization and quality control, remote sensing methods and review of agency monitoring programs.

These research areas can be organized or grouped in a somewhat different way to better understand their rationale. They can be revolved into three groups:

1. *water management*, which includes the sociological, political, attitudinal and other factors related to large scale management of water resources including quality (systems approach)
2. *hardware*, which includes improvements in present mechanical equipment and investigation of alternate or combined systems
3. *new methodology*, which includes entirely new approaches such as sewage distillation and dehydration.

Major areas of research that should be given future attention include:

1. understanding of the organic and inorganic processes whereby heavy metals are mobilized

in the ecosphere, their passage through the
food chain and their concentration or fixation
in life forms and sediment
2. the economic and technical feasibility of
removing heavy and/or toxic metals from waste
streams
3. the effects on life forms and the removal of
organic chemicals including pesticides like
DDT or PCB
4. the persistence and longevity of biological
pathogens and methods of removal
5. methods of improving processing technology
to prevent the introduction of these materials
into the waste stream and/or substitution for
their use in industrial processes.

REFERENCES

1.   Wollman, N. and G. W. Bonem. *The Outlook for Water*
(Baltimore, Md.: Resources for the Future, Inc., Johns
Hopkins Press, 1971).

SECTION II

WATER POLLUTION

5. INTRODUCTION

Water is one of the most valuable substances we have. Three-fifths of the surface of the earth is covered by water, mostly encompassing the oceans of the world. Rivers, lakes, and ground water supply most of our water needs even though they constitute a very minor portion of the hydrosphere. These sources of water are rapidly becoming polluted with the spoils of human progress. Many of the extractive industries are guilty of some water pollution.

Coal mining is one of the prime contributors to the deterioration of our streams. The Appalachian Regional Commission (1969)[1] has reported that in the U.S. alone more than 10,500 miles of streams have been polluted by mine drainage.

Steel mills, copper smelters, cement plants, wrought iron works, flotation operations and oil refineries, to name just a few, contribute many pollutants to our waterways. Table VI lists some of the waste characteristics of some of our extractive industries.

Most of the extractive industries consume vast amounts of water. Jones & Laughlin's Pittsburgh Works uses 153 billion gallons of water per year, a quantity comparable to the amount of water consumed by all other users of water in the City of Pittsburgh.[2]

The cost of pollution control required to remove contaminants to conform to federal, state and county regulations is tremendous. Large corporations are having to spend a larger share of capital investment each year for pollution control equipment, which is usually rather bulky and expensive to operate. A usual rule-of-thumb is that the operating cost of a water pollution control device averages 10% per year of its purchase price.

Because of the high investment required and the state of flux of environmental guidelines, many corporations are fighting to delay the installation of

43

Table VI

Waste Characteristics of Some Extractive Industries

Industry	Waste Characteristics
Salt mining	high in suspended solids, high salts, corrosive
Coal mining	fine suspended coal, high total dissolved solids, acid mine drainage, trace metals
Integrated iron & steel industry	high total dissolved solids, high iron, high acidity, high concentration of many metallic salts, pickle liquor, mill scale, phenols, alcohols, cyanides, tars, oils, heated water
Smelters	high sulfur dioxide, high cyanides, phenols, and finely-divided suspended matter
Electroplating	high in salts of copper, zinc, chromium, and nickel, high in cyanides
Oil refineries	high in organics, acids, alkalies, phenols, and some trace metals, high in oil-coated solids
Flotation plants	high in mercaptans, cyanides, arsenic, and many toxic metals

control devices. They resort to requesting variances from county or state hearing boards to allow time for further study and permit evaluation of the cost-benefit ratios of various alternatives. Such an attitude coupled with occasional threats of shutdowns has given the extractive industries a rather bad reputation as anti-environmentalists. Labor unions such as the United Steelworkers of America have entered the picture by providing information to union members regarding "environmental blackmail" and possible "threats of job losses" because of pollution control.

It has indeed been established that those companies that are most likely to be unable to meet environmental control guidelines are those presently suffering marketplace pressure and already requiring major modernization. Sometimes in asking for variances, companies are simply asking covertly for extensions to allow them to operate a plant already slated for eventual shutdown.

Pollution control in the extractive industries is very complicated because of the number of processes and operations required to transform rocks and minerals into valuable products. The vast number of operations required (Table VII) in most of the extractive industries result in a large number of vastly different waste streams. Some of the resulting waste products can be dumped together and treated in one final treatment step prior to eventual disposal, but others are so noxious that they must be kept separate and then be carefully packed prior to disposal so they never come in contact with the environment.

Table VII

Some Processes in Steelmaking and Petroleum Refining

Steelmaking	*Petroleum Industry*
Coking	Storage and transport of crude
Blast furnace	Desalting
Steelmaking furnace	Crude fractionation
Casting	Solvent dewaxing
Rolling	Wax finishing
Pickling	Grease making
Cleaning	Lubricating oil finishing
Galvanizing	Hydrotreating
Tinplating	Catalytic cracking
Polishing	Thermal cracking
Metal coating	Drying
Sintering	Storage of finished products
Scarfing	Piping

Because of the complexity of water pollution abatement in the extractive industries, it is imperative that each individual industry conduct a feasibility study of needed in-plant abatement (Table VIII). Such a feasibility study can then be used to formulate a pollution abatement plan (Figure 5) and a timetable for compliance with environmental guidelines.

The challenge of meeting the mineral (oil, gas, and water included) needs of the world in future decades while at the same time minimizing environmental degradation must be met by operating companies. Especially important problems that must be met include:

1. reduction of water pollution from integrated
   metallurgical operations
2. prevention of acid mine drainage from operating
   and abandoned coal mines
3. reclamation of dumps and stripped lands
4. elimination of fluoride emissions from the
   aluminum and phosphate industries
5. prevention of groundwater and runoff
   contamination
6. prevention of oil pollution from refinery
   operations, oil spills, and tanker collisions.

Table VIII

Pollution Control Feasibility Analysis

1. Definition of Problem Areas

   a. Type and extent of pollution problems in each
      area
   b. Evaluation of process flow paths
   c. Evaluation of individual waste toxicity
   d. Evaluation of waste stream flow path

2. Design of Waste Handling System

   a. Separation of waste into streams depending on
      waste composition, concentration, and
      toxicity
   b. Appropriate treatment applied to the waste load
      prior to eventual discharge or reuse. Treatment
      must be such that any discharges meet
      environmental guidelines

3. Technology Improvement

   a. Evaluation of waste load changes resulting from
      process improvements
   b. Evaluation of waste discharges resulting from
      improved pollution control equipment

4. Economic Analysis

   a. Cost-benefit evaluation of various alternatives
      for pollution abatement
   b. Evaluation of capital costs required for various
      levels of pollution control
   c. Evaluation of operating costs required for
      various levels of pollution control
   d. Evaluation of alternative means of financing
      pollution control equipment

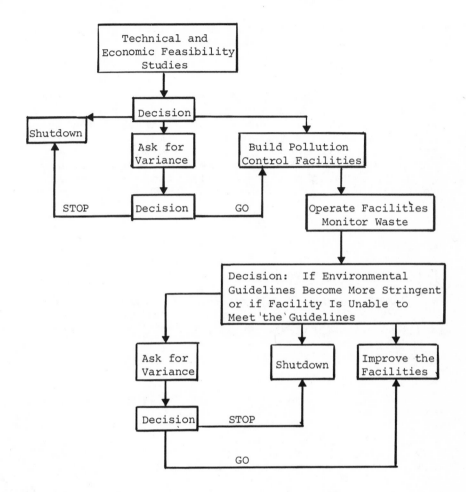

Figure 5.   Steps in planning pollution abatement.

Operating companies must be required to comply with fair environmental standards set by appropriate government agencies. The laws must be written to insure ease of resource extraction with minimal environmental degradation. Appropriate government agencies should periodically police the extractive industries to insure proper compliance with environmental guidelines. Fines should be strong enough to insure compliance. Variances from any regulation should be kept at a minimal level. Tax incentives to industry could be devised based on:

1. extent of reclamation activities
2. amount of environmental research
3. record of compliance above set environmental standards.

The papers in this section discuss the problems involved in mining and metallurgical operations, as well as the overall levels of industrial pollution and new techniques for regional water pollution monitoring.   In subsequent sections we will consider the problems of disposal of refinery effluents and metal processing plants as well as the subject of reclamation.

REFERENCES

1.   Appalachian Regional Commission.   "Acid Mine Drainage in Appalachia" (1969).
2   Jones & Laughlin Steel Company.   "J & L and the Environment: A Review of the Corporation's Activities in the Field of Air and Water Quality Control," 3rd ed. (May, 1973).

6. COAL MINING AND ITS EFFECT ON WATER QUALITY

Moid U. Ahmad
Department of Geology
Ohio University
Athens, Ohio

ABSTRACT

Coal mining has seriously degraded the water quality
of streams and aquifers in Appalachia and other parts of
the United States. In Appalachia about 6,000 tons of
sulfuric acid are produced every day as a result of
pyrite oxidation. A typical acid discharge may have pH
levels of 3.5-6.6, acidity content of 0-1000 ppm, sulfate
content of 500-1,000 ppm, and an iron content of 0-500
ppm. The flow pattern of surface and subsurface waters
is being seriously disturbed in over 11 million acres of
land mined for coal in the United States. Large areas
of stripped land cannot supply good quality water. Rural
water supplies in coal mining districts are being seriously
threatened by stripping.

One of the strip mines that has been studied in detail
is the Sheban Mine covering 79 acres near Youngstown,
Ohio. This mine produces 60 million gallons of highly
acidic water per year, with the drainage seriously
affecting the drinking water supply of Youngstown.

Oxidation of pyrite in underground mines results in
the formation of hundreds of unnatural springs discharging
highly acidic water into our streams. No effective method
has yet been devised to stop acid production.

INTRODUCTION

    Coal, the nation's most abundant fuel resource, provides over 18% of our energy needs and is used to generate 44% of our electric power. Coal deposits underlie 350,000 mi^2 of the total area of the United States. Total U.S. coal production in 1971 was 552 million tons, half of which was surface-mined. Sixty-eight per cent of U.S. coal production is in Appalachia, an area that accounts for 56% of total U.S. surface-mined coal.

Acid Mine Drainage

    The oxidation of pyritic materials accounts for the acidity of mine drainage. This oxidation can be described by the following reactions:

$$FeS_2(s) + \frac{7}{2}O_2 + H_2O \rightarrow 2SO_4^{-2} + Fe^{+2} + 2H^+ \qquad (1)$$

$$Fe^{+2} + \frac{1}{4}O_2 + H^+ \rightarrow Fe^{+3} + \frac{1}{2}H_2O \qquad (2)$$

$$Fe^{+3} + 3H_2O \rightarrow Fe(OH)_3 + 3H^+ \qquad (3)$$

Often, the three equations are added, giving the overall stoichiometric relationship:

$$FeS_2(s) + \frac{15}{4}O_2 + \frac{7}{2}H_2O \rightarrow 2H_2SO_4 + Fe(OH)_3$$

Smith and Shumate[1] have described the rates of pyrite oxidation and acid production in the field. Equation (1) represents the oxidation of crystalline pyrite by oxygen, which is the critical reaction in the scheme. The reaction sites are at the pyrite crystal surface exposed to oxygen. Equations (2) and (3) represent the oxidation of ferrous iron to ferric iron and subsequent precipitation of ferric iron as ferric hydroxide, which gives the familiar red color observed in acid streams.

    For fine-grained pyrite, the rate of the reaction represented by Equation (1) is a function of the oxygen concentration, temperature, degree of surface saturation by water, and pH of the solution in contact with the pyrite. It is important to note that the conditions at the pyrite surface control the rate of the oxidation reaction.

These equations are useful for illustrating the
production of acidity in mine drainage. The impression given is that only iron compounds and sulfuric
acid occur in mine drainage. Secondary reactions
take place between ferrous sulfate, sulfuric acid,
and compounds in clays, shales, limestones, sand-
stones and other strata associated with the coal
beds. Mine drainage can be regarded as a solution
of mixed salts, mostly sulfates. Recent studies
indicate that mine drainage waters contain toxic
elements.

STRIP MINING

Surface mining of coal greatly disturbs the
natural and physical characteristics of the land on
which it occurs. Exposed pyritic materials in the
highwall or scattered on the spoil banks will
oxidize and produce acid drainage from the mine.
This accounts for about 25% of the total acid
production in Appalachia.
A far more serious source of pollution is that
of silt created by the erosion of the spoil banks
in strip mines. Erosion may occur during the early
periods of mining when the spoil banks are very
loose and subject to slipping.

Hydrological Studies of Strip Mines

A number of studies have been conducted by
various researchers regarding some aspects of the
hydrogeology of strip mines. During the three year
study period in Pike County, Indiana, Corbett[2] reported that the spoil bank aquifers materially
reduce major flood flows; on the other hand, these
aquifers increase flow during extended dry periods
and during summer and fall seasons when the stream
flow is below lowland flooding. Similar findings
are reported in the report of Piedmont Reservoir
Watershed by the Department of Natural Resources.[3]
However, Neely[4] reported that lands affected by
strip mining (even though reclaimed) are losing
their nutrients at alarming rates in the Piedmont
Reservoir.
An exhaustive study of the influences of strip
mining on the hydrological environment of parts of
Beaver Creek Basin, Kentucky was conducted by
Collier *et al.*[5] They reported that the effects of

strip mining on the runoff characteristics are not readily differentiated from other factors in the hydrological environment, which also affect the runoff chemical quality, increase the sediment yield of the basin, alter the forest, and the aquatic life of the streams. Sternberg[6] constructed a mathematical model representing a strip-mined area and obtained solutions for the changes in ground water elevation and ground water flow that would occur in response to a uniform rate of deep percolation over the spoil banks. Shumate and Brant[7] stated that "it is unlikely that material buried several feet or more beneath the surface can undergo significant oxidation because of the restriction of oxygen diffusion to these depths, and the reaction zone is most probably restricted to the outer layer of the pile."

Lovering[8] showed that chalcopyrite and pyrite at a depth of 600 to 700 feet in Arizona are oxidizing and producing temperature anomalies at the surface. Collier *et al.*[5] reported that one spoil bank in which pyritic material was deeply buried did produce acid water after two years. Temperature surveys of strip mines[9] have indicated that pyrite oxidation may be occurring at a depth of four feet.

Extensive studies have not been done to determine the effects of mine drainage on ground water.

Sheban Strip Mine

The Sheban Mine (79 acres) is a typical coal mine.[9] The mine is located in the N.E. quarter of Goshen Twp, Mahoning County, Ohio. Drainage leaves the affected area, flows east under the Stratton Road, thence into the upper reach of Meander Creek tributary through Ponderosa Lake and eventually into Meander Reservoir, which supplies water to Youngstown, Ohio. Approximately 24 acres of the watershed consist of acid spoil having 5% vegetation covering the surface. South of the lake is undisturbed land. About 55 acres of the spoil bank material northwest of the lake was leveled, heavily limed, and then covered with manure to a depth of six inches or more. Following decomposition of the manure, the land was seeded and presently is a good pasture land. Effluent from the western half seems to be more acidic than the effluent discharged from the entire mine. The analysis of water at the Stratton Road sampling station* reveals the following composition:

*Data supplied by Chief Engineer, Mahoning Valley Sanitary Dist.

constituent	ppm
Fe	142
Cl	15
F	11
Cd	0.025
Cu	0.08
Mn	135
Ni	2.5
Zn	2.6
Al	100
Ca	520
Mg	550
Na	12
K	4
acidity	323
hardness	5000
SO₄	2167
pH	3.0

The Chief Engineer of the Mahoning Valley Sanitary District has noted that the Sheban Mine supplies about 60 million gallons per year of highly acidic water to the Meander Reservoir, which is the source of drinking water for Youngstown, Ohio.

ACID MINE DRAINAGE FROM UNDERGROUND MINES

In places where overlying rock materials are thick, underground mining must be employed. Where the coal crops out at the land surface, a drift mine entry is constructed, and mining is advanced into the coal bed. It is this type of mining above the surface drainage that is producing 75% of the total acid load in Appalachia. There are several thousand underground sources producing the acid discharge either continuously or intermittently. Of the sources that have been inventoried, inactive underground mines contribute 52% of the acid and active ones 19%.

Mine Environment and Acid Production

During the mining process a great deal of coal and pyrite is left in the mine. The mechanism of acid production has already been described. The vehicle of transport for the oxidation product is flowing ground water. In another situation, it might be the intermittent percolation of rain water down through the overburden. Oxidation products may either diffuse along water films to a region of active ground water flow or, because of the hydro-scopic nature of these salts, they may condense water from the surrounding air to promote sufficient seepage to carry products to regions of active ground water flow.

When intermittent flow of percolating rain water or a rising ground water table approaches or inter-sects the region of product storage, there will be a surge of acid products from the system. This is regularly noticed in many instances in the spring in the increase of ground water flow. Even though oxidation sites may be flooded out during such a period and the total rate of pyrite oxidation may be at a minimum, the rate of transport of products out of the system might be at its maximum. The water entering the mine is generally of good quality and comes out acidic. If limestone is present, the water is neutralized and may have a high pH. Such waters are high in iron and sulfates.

MAGNITUDE OF THE PROBLEM

Ahmad[10] has shown that strip mines may be mapped using ERTS-A pictures. Figure 6 is a picture of SE-Ohio showing that the hydrological and soil characteristics are remarkably different from the surrounding land. Spaulding and Ogden[11] have re-ported that approximately 2 million acres of land are affected. Wildlife habitat is affected in 13,000 miles of streams (13,970 surface acres), 281 natural lakes (103,630 surface acres), 168 reservoirs and impoundments (41,516 surface acres). Significant damage to the streams outside Appalachia is noted below:

	Stream Miles	Surface Acres	Reservoirs and Impoundment Surface Acres
Missouri	330	2,550	6,000
Colorado	880	1,930	600

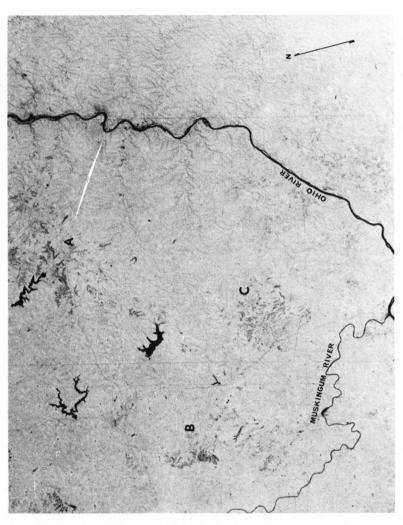

Figure 6. Blow-up of ERTS—A picture showing SE-central Ohio, approx. scale 1:500,000. Letters indicate major stripped areas.

REFERENCES

1. Smith, E. E. and K. S. Shumate. "Rate of Pyrite Oxidation and Acid Production Rate in the Field," Proceedings Acid Mine Drainage Workshop, Ohio University (1971).
2. Corbett, D. M. "Acid Mine-Drainage Problem of the Patoka River Watershed," Rep. of Inv. #4, Indiana University, Bloomington, Indiana (1969).
3. Department of Natural Resources. Annual Report "Stream Quality Surveillance Program—Piedmont Reservoir Watershed, Belmont and Guernsey Counties, Ohio," (1969).
4. Neely, J. C. "The Effects of Strip Mining on a Natural System, A Water Quality Study of Piedmont Lake," Case-Western Reserve University, Cleveland, Ohio (1970).
5. Collier, C. R., *et al.* "Influence of Strip Mining on the Hydrological Environment of Parts of Beaver Creek Basin, Kentucky," Geological Survey Professional Paper 427-C (1970).
6. Sternberg, Y. M. and F. A. Allen. "Hydrology of Surface Mining—A Case Study," *Water Resources Research, 4(2),* (1968).
7. Shumate, K. S. and R. A. Brant. "Acid Mine Drainage Formation and Abatement," WPCR Series, EPA (1971).
8. Lovering, T. S. "Geothermal Gradients, Recent Climatic Changes and Rate of Sulfide Oxidation in the San Manuel District, Arizona," *Econ. Geol. XLIII(1),* (1948).
9. Ahmad, M. U. "A Hydrological Approach to Control Acid Mine Pollution," Proceedings Acid Mine Drainage Workshop, Ohio University (1971).
10. Ahmad, M. U., *et al.* "Mapping of Spoil Banks Using Earth Resources Technology Satellite—A Pictures," Second Annual Remote Sensing of Earth's Resources Conference, University of Tennessee. (1973).
11. Spaulding, J. and R. D. Ogden. "Effects of Surface Mining on the Fish and Wildlife Resources of the United States," U.S. Dept. of Interior (1968).

7. ACID PRODUCTION IN MINE DRAINAGE SYSTEMS

Michael J. Smith
Department of Chemistry
Wright State University
Dayton, Ohio

ABSTRACT

An effective method for the prevention of acid production from mine drainage systems can be developed only if an understanding of the kinetics and reactions responsible for acid mine drainage are available. A believable reaction sequence has been described[1] for the dissolution of pyrite in the presence of the appropriate microorganisms and dissolved oxygen, yet little is known about the specific chemical interactions that take place in mine drainage systems. A discussion of the chemistry of such systems is presented in this paper.

INTRODUCTION

A number of bacteria can live autotrophically on the energy released by the oxidation of inorganic iron or sulfur compounds. Within this group is a subgroup of *strict aerobes*, which comprise the genus *Thiobacillus*. Two species of this genus, *Thiobacillus thiooxidans* and *Thiobacillus ferrooxidans*, (sometimes designated *Ferrobacillus ferrooxidans*), have been described as important in the formation of acid in mine drainage.[2] Acidic conditions result from the release of hydrogen ions during the oxidation of the iron sulfides, pyrite and marcasite, and the subsequent formation of sulfuric acid. Some strains of *T. thiooxidans* are remarkably resistant to acid and have been found active even at pH 0. Acid tolerances for these bacteria range typically from pH 5.2 to 1.5.[3]

57

The "ideal" thiobacillus is able to oxidize virtually all stable reduced sulfur species including S^{-2}, S^0, $S_2O_3^{-2}$, and $S_4O_6^{-2}$. The word ideal is used here because apparently not all strains of thiobacilli are able to carry out all these oxidations under all conditions.

The inorganic transformations brought about by these microorganisms are of considerable economic importance to man. The clogging of iron pipes used to convey sulfurous water from coal deposits is a direct result of action by the thiobacilli, which oxidize ferrous iron to ferric iron for a primary energy-generating process in the fixation of CO_2. Because very little energy is gained in the oxidation, a small number of bacteria can be responsible for oxidizing large amounts of ferrous iron.

Apart from their activities in mine water, the thiobacilli are economically important in other ways. Air comprised of various reduced sulfur species such as H_2S and SO_2 is apt to be found in urban industrial areas. As a consequence, the water film on stonework is likely to contain sulfur species that can be oxidized by thiobacilli. The thiobacilli have equally disastrous effects on concrete, and at least one study, by Parker and Prisk,[4] was completed using a bacteria isolated from corroded concrete. Thiobacilli are also present in acid sulfur springs and are responsible for most of the iron and sulfur precipitated at very low pH values. These acid waters are toxic to fish and plants and as a result are undesirable.

Little is known about the abilities of various strains of thiobacilli to oxidize various species except that all are able to oxidize thiosulfate. Conflicting results concerning the ability of these organisms to oxidize H_2S are reported in the literature.[4,5] Parker and Prisk concluded that only two of seven strains of thiobacilli investigated could do such, but the H_2S in their experiment was supplied at the fairly high rate of 200 ppm in the air used to aerate the cultures. At this level H_2S may have been toxic to some species that could have oxidized it at lower concentrations. Later, Beck[5] investigated an iron-oxidizing bacterium isolated from acid-leaching water that was morphologically identical to one of the sulfide-oxidizing bacteria investigated by Parker and Prisk and found that it oxidized Fe^{+2} and sulfur mixtures more rapidly than when only one or the other was present. No attempt was made to use sulfide as an energy source. Because of inconsistencies like this in the literature, the

known facts concerning iron and sulfur species oxidation are indeed few.[1] It should also be noted
that the interpretation of the literature concerning
the thiobacilli is often hampered by either an in-
complete or inadequate description of the products
of the oxidation reactions that these organisms
cause.

Because of the great diversity of conflicting
experimental data presented in the literature and
for the sake of brevity, the remainder of this discussion will be concerned with the chemical aspects
of the acid mine drainage problem and the two
organisms felt to be most important in the formation
of acid mine waters, *T. thiooxidans* and *T. ferrooxidans*
A short description of the characteristics of the
two organisms and a discussion of the microbially
mediated oxidation reactions they are able to utilize
as energy sources will be followed by a discussion
of the reaction sequence whereby the oxidation of
iron and sulfur species is thought to occur. In
order to emphasize the behavior of iron and sulfur
in natural waters, their equilibrium and kinetic
relationships in the absence of microorganisms will
be discussed and their relative importance evaluated.
Because the thiobacilli, like other organisms, affect
their environment as much as their environment affects
them, the information presented will be integrated
in an attempt to reach some conclusion concerning
the relationship between the distribution of
thiobacilli, their environment, and the chemistry
of aqueous iron and sulfur species.

THE NATURE OF PYRITIC MATERIALS
FOUND IN COAL DEPOSITS

Iron disulfide exists in one of two crystalline
forms—pyrite or marcasite. In a recent study, of
30 samples collected from representative coal forma-
tions in the Eastern United States, all but two were
found to contain pyrite as the major iron sulfide
constituent. In the majority of samples marcasite
was either absent or present in amounts of less than
5%.[6]

The shape and size of naturally occurring
pyritic materials is used as a method of classifica-
tion. Very small-grained, brassy-grey, flattened
spherical masses are distinguished as "sulfur balls."
Pyrite comprises 90 to 98% of the sulfur ball with
coal making up the remaining percentage. The

porosity and large surface area of sulfur ball materials makes them much more reactive than other more compactly deposited pyritic materials.

The oxidation products found on the surfaces of pyrite-containing materials are hydrates of $FeSO_4$, usually $FeSO_4 \cdot 7H_2O$. After dilution by streams running from mined areas, Fe(II) salts are converted to $Fe(OH)_3 \cdot (H_2O)_x$ which is deposited in stream beds and surrounding areas. These deposits are termed "yellow boy" in the mining industry.

DESCRIPTION AND CHARACTERISTICS OF *T. THIOOXIDANS* AND *T. FERROOXIDANS*

Thiobacillus ferrooxidans

T. ferrooxidans was first isolated by Colmer and Hinkle[7] from acid mine waters. Little is known about its distribution elsewhere. These bacteria are characterized as rod-shaped (0.8-1.0 x 0.4 μ) with a single flagellum. They are one of the principal organisms participating in the oxidation of heavy metal sulfides. Colmer, Temple, and Hinkle[8] reported that in addition to oxidizing ferrous salts, this species also derives its energy from reduced sulfur compounds. In their experiments, the bacteria oxidized thiosulfate with precipitation of elemental sulfur and the accumulation of sulfates. Repeated subculturing of the organism on an iron-containing medium results in the loss of ability to oxidize thiosulfate, though thiosulfate grown cells are still capable of oxidizing ferrous salts.[9]

The existence of this species was questioned by Leathen *et al.*[10] who could not find a single bacterium that was capable of oxidizing both thiosulfate and ferrous iron. Hence a new name for the organism, *Ferrobacillus ferrooxidans*, was suggested; later, however, Beck[5] confirmed that Colmer's original findings were correct. In addition, he found that the bacteria could oxidize sulfur as well as thiosulfate. An interesting feature of this organism is its remarkable resistance to high concentrations of copper and other heavy metals.[11] The bacterium can be grown in a medium containing up to 15 g Cu^{+2} per liter (0.24 M $CuSO_4$).

Thiobacillus thiooxidans

T. *thiooxidans* utilizes the energy derived from the oxidation of reduced sulfur species for its biosynthetic processes in which CO_2 species function as their sole source of carbon. These organisms neither use organic materials nor is their growth suppressed by them. In contrast to other thiobacilli, most of which prefer a nearly neutral environment, *Thiobacillus thiooxidans* grows best below pH 5 and has been known to grow in 10% (1 M) H_2SO_4.[11] Aside from its acid tolerance, this bacterium is primarily distinguished by its ability to oxidize elemental sulfur at a rate comparable to its oxidation of thiosulfate. T. *ferrooxidans* can be differentiated from the latter by its inability to oxidize sulfur rapidly.[9] T. *thiooxidans* was originally isolated from soil enriched with elemental sulfur and is commonly found along with T. *ferrooxidans* in sulfide ore deposits. Marine muds are also a reliable source of T. *thiooxidans*. Their occurrence has been attributed to the presence of H_2S produced by sulfate reducing bacteria.

MICROBIAL OXIDATION OF IRON AND SULFUR SPECIES BY *T. THIOOXIDANS* AND *T. FERROOXIDANS*

While it has been known that the oxidation of sulfide, sulfur, thiosulfate, and iron species function as energy sources for the thiobacilli, the intermediate species formed during the various oxidative steps are still largely unknown. Because of the economic impact of the acid mine drainage problem, the thiobacilli have been studied largely from the standpoint of their effects on the oxidation of pyritic materials. Such systems are inherently complicated. As a consequence, the direct observation of the exact nature of the oxidation reactions the thiobacilli can catalyze in simple laboratory systems has not been adequately made.

Thiobacillus ferrooxidans

T. *ferrooxidans* can rapidly oxidize Fe^{+2} under aerobic conditions in the absence of reduced sulfur species in a sulfuric acid-containing medium:[5,8]

$$Fe^{+2} + H^+ + 1/4\ O_2 \rightarrow Fe^{+3} + 1/2\ H_2O \qquad (1)$$

The uptake of oxygen is stoichiometric with the above reaction and CO_2 fixation is observed to cease abruptly with depletion of Fe^{+2}. It has also been demonstrated that the rate of oxidation of Fe^{+2} is proportional to the number of bacterial cells present. Cell suspensions of this organism are able to oxidize sulfur in the absence of Fe^{+2},[5] but because of variations of the S^0 actually in contact with the cells, a stoichiometric relationship between initial sulfur content and oxygen consumed has not been established. The rate of oxygen uptake in media containing both S^0 and Fe^{+2} has been observed to be double that in media containing sulfur alone. No explanation has been given for this rather interesting phenomenon. In addition, the efficiency of carbon fixation was found to range from 0.21 to 0.30 moles of CO_2 fixed per mole of O_2 consumed during Fe(II) oxidation and from 0.015 to 0.025 moles of CO_2 fixed per mole of O_2 consumed during S^0 oxidation. On a mixed substrate, the efficiency of carbon fixation increased nearly three-fold over S^0 alone to 0.081 mole CO_2 per mole O_2. Growth in acid thiosulfate media was very slow[5] and perhaps did not occur at all since the formation of colloidal sulfur obscured turbidity measurements used to determine the number of cells present.

Leathen, Braley, and McIntyre[10] have reported that *T. ferrooxidans* enhances acid formation from marcasite, and "sulfur balls" (a type of porous pyritic concretion) but not from massive pyrite. Later Temple and Delchamps[1] found that *T. ferrooxidans* could also increase acid formation from pyrite of small particle size. Several other studies have revealed the ability of the organism to oxidize the sulfides of Fe, Cu, Sb, Mo, etc. Experiments performed by Brynner, *et al.*[11] indicate the biological character of pyrite oxidation. It is also known that the oxidation of mixed sulfide ores is markedly enhanced by the presence of *T. ferrooxidans*.

Thiobacillus thiooxidans

The oxidation of reduced sulfur species by *T. thiooxidans* is in many cases open to question. In the absence of other data on the subject, it remains unclear whether this organism is capable of oxidizing H_2S. Early investigations of this oxidation indicated that H_2S could not be utilized, but later work[4] indicates that it is oxidized to sulfuric acid when maintained in a liquid mineral media incubated in air containing 200 ppm H_2S. The accumulation of a small amount of elemental sulfur toward the end of their experiment suggests that sulfur may have been formed initially by purely chemical means and was subsequently oxidized to H_2SO_4 by the organism:

$$1/2 \ O_2 + H_2S \rightarrow S^0 + H_2O \qquad (2)$$

$$3/2 \ O_2 + H_2O + S^0 \xrightarrow{\text{Thiooxidans}} SO_4^{-2} + 2H^+ \qquad (3)$$

It has been suggested that sulfur is the main source of energy for *T. thiooxidans* because of its ability to rapidly oxidize elemental sulfur directly to H_2SO_4.[4,12] Rhombic sulfur, sulfur precipitated from sulfides treated with mineral acids, and amorphous sulfur (from melted ground sulfur) are all rapidly oxidized. It has been experimentally established that the rate of this oxidation is related to the sulfur particle diameter, decreasing with increasing size up to 1.3 mm.

Thiosulfate at low concentrations is quantitatively converted to sulfate without intermediate product. In media containing high concentrations (greater than 1%) of $S_2O_3^{-2}$, a small amount $S_4O_6^{-2}$ (tetrathionate) is formed as an intermediate product although the rate of acid formation is slower than that with S^0:

$$S_2O_3^{-2} + 2O_2 + H_2O \rightarrow 2SO_4^{-2} + 2H^+ \qquad (4)$$

$$2H^+ + 1/2 \ O_2 + 2S_2O_3^{-2} \rightarrow S_4O_6^{-2} + H_2O \qquad (5)$$

Temple and Delchamps[1] found that *T. thiooxidans* was unable to oxidize pyrite but that both marcasite and "sulfur balls" were oxidized by this organism under aerobic conditions. Mixtures of *T. thiooxidans* and *T. ferrooxidans* gave the same rate of oxidation as *T. ferrooxidans* alone. Substrates consisting of covellite (CuS), massive pyrite, and MoS were not oxidized by *T. thiooxidans* in later experiments conducted by Sokolova.[13]

In summary, it appears that the primary energy-producing reaction for *T. thiooxidans* is the oxidation of elemental sulfur directly to SO_4^{-2} without intermediate product formation. It also appears that *T. ferrooxidans* is capable of oxidizing Fe^{+2} to Fe^{+3} very rapidly. The two organisms together are capable of oxidizing pyritic material at about the same rate as *T. ferrooxidans* alone. A summary of the transformations that are catalyzed by these organisms is given in Table IX.

KINETIC AND EQUILIBRIUM RELATIONSHIPS
OF Fe(II) AND Fe(III)
THE CHEMISTRY OF THE FeS_2 SYSTEM

The iron and sulfur species that exist in natural waters are determined by equilibrium relationships provided that the kinetic relationships governing the species are such that equilibrium is attained. The deviations from the behavior predicted by thermodynamic means are usually attributed to conditions where the kinetics of the chemical reactions involved are limiting. Typical compositions of acid mine waters are given in Table X. Acid mine waters from deep mines commonly yield E_h values between +0.08 v and -0.04 v. These E_h values have been measured at corresponding pH values of 3.95 and 5.41.[2] These conditions are typical of those for pyritic materials. Another deep mine had E_h values of about +0.55 at pH values of 3.5, which correspond to conditions where large amounts of Fe^{+2} are present. Strip mines have higher E_h values because they are exposed to the atmosphere.

Under the conditions and pH ranges that normally occur in mine drainage waters, the only form of Fe(II) of importance is Fe^{+2}. $FeCO_3(s)$, $FeHCO_3^+$, and $FeSO_4$ have little or no tendency to form.[13] Complexation is important, however, in determining the Fe(III) species present. At pH values greater than 1, S(VI) is present primarily as SO_4^{-2} or HSO_4^- as is indicated by the magnitude of $K_{a,2}$ for HSO_4^-. At a total sulfate concentration of 1×10^{-2} M, the extent of competition for Fe^{+3} by the coordinating ligands OH^- and SO_4^{-2} is given in Table XI. At the concentrations of SO_4^{-2} commonly encountered in acid mines, sulfate complex formation is significant. Ferric hydroxide precipitation begins above a pH of 3 (assuming a maximum $[Fe^{+3}] = 2 \times 10^{-3}$ M). Hence, the solubility of ferric iron is limited by its rather insoluble hydrated oxide even at the very low pH values occurring in acid mine water.

Table IX

Summary of the Characteristics of the Energy Yielding Chemical Oxidations by Two Thiobacilli Important in the Formation of Acid Mine Waters

Organism	Cell Dimensions	Source of Energy	Reference	Relative Rate of Reaction for Species Indicated
T. thiooxidans pH Range: ~0→6	1 × 0.5 μ	$H_2S \rightleftharpoons SO_4^{-2} + H^+$	4	slow
		* $S^0 \rightarrow SO_4^{-2} + H^+$ *	4,5	very rapid
		$S_2O_3^{-2}$ <1% $S_4O_6^{-2} + SO_4^{-2} + H^+$	4	slower than S^0
		$S_2O_3^{-2}$ >1% $SO_4^{-2} + H^+$	4	
		$FeS_2 \rightarrow H^+ + ?$		pyrite = 0, marcasite > "sulfur balls"
T. ferrooxidans pH Range: <2→4.5	0.8-1 × 0.4-0.7 μ	* $Fe^{+2} O_2, H^+ \rightarrow Fe^{+3}$ * SO_4^{-2}	5	very rapid
		$S^0 \rightarrow SO_4^{-2} + H^+$	5	slow
		$S^0 Fe^{+2} \rightarrow SO_4^{-2} + H^+$	5	faster than with S^0 only
		$S_2O_3^{-2} \rightleftharpoons S^0 + SO_4^{-2}$	7,8	slower than with S^0, rate = 0 if acclimated to Fe^{+2} first
		$FeS_2 \rightarrow Fe^{+3} + SO_4^{-2} + H^+$	5	marcasite > pyrite > "sulfur balls"

Table X

Composition of Mine Drainage Waters
Before Dilution by Streams

Parameter or Ion	Common Range	Typical Value
pH	2.0 - 6.0	3
Fe^{+2} (ppm)	10 - 2,000	100 (2 x 10^{-3} M)
Fe^{+3} (ppm)	0 - 100	0
SO_4^{-2} (ppm)	100 - 2,000	1,000 (1 x 10^{-2} M)
HCO_3^- (ppm)	0 - 200	0
Ca^{+2} (ppm)	10 - 1,000	200 (5 x 10^{-3} M)
Al^{+3} (ppm)	0 - 150	50 (2 x 10^{-3} M)
ionic strength	---	0.3
acidity (eq/l)	---	2 x 10^{-3}
O_2 (atm)	0.20 - 10^{-83}	10^{-60} (deep mines)

Table XI

Extent of Coordination of Fe(III) by
OH^- and SO_4^{-2} at Low pH Values

pH	$[FeSO_4^+]/[Fe^{+3}]$*	$[Fe(OH)^{+2}]/[Fe^{+3}]$*
4	5.3	15.0
3	4.9	1.5
2	3.0	0.15
1	0.63	0.015

*calculated from:

$$\frac{[FeSO_4^+]}{[Fe^{+3}]} = K_{f,FeSO_4^+} \times C_{Sulfate} \times \frac{K_{a2,H_2SO_4}}{[H^+] + K_{a2,H_2SO_4}}$$

$$\frac{[Fe(OH)^{+2}]}{[Fe^{+3}]} = K_{f,Fe(OH)^{+2}} \times [OH^-]$$

where

$$C_{Sulfate} = 1.0 \times 10^{-2} M$$

$$K_{f,FeSO_4^+} = 5.3 \times 10^2$$

$$K_{a2,H_2SO_4} = 1.3 \times 10^{-2}$$

$$K_{f,Fe(OH)^+} = 1.5 \times 10^{11}$$

The rate of oxidation of Fe(II) by oxygen has been found to follow a first order relationship for Fe^{+2} and O_2 and a second order relationship for OH^- above a pH of 4.5. The reaction proceeds rapidly at high pH's with a half-time of 4.0 minutes at pH 7.0 at a P_{O_2} of 0.2 atm at 25° C. At a pH of 4.5 the half-time increases to approximately 300 days. At pH values less than 2, the rate of reaction is independent of pH with a half-time of 2,000 days. A plot of oxygenation rate of Fe^{+2} as a function of pH is shown in Figure 7. Note that the variation in rate with pH coincides with the pH at which $Fe(OH)_3$ begins to precipitate.

The following series of reactions have been proposed for the oxidation of pyrite at pH values greater than 3 after exposure to O_2 during mining operations or in a mine drainage stream bed:

$$FeS_2 + \frac{7}{2} O_2 + H_2O \rightarrow Fe^{+2} + 2SO_4^{-2} + 2H^+ \tag{6}$$

$$Fe^{+2} + \frac{1}{4} O_2 + H^+ \rightarrow Fe^{+3} + \frac{1}{2} H_2O \tag{7}$$

$$Fe^{+3} + 3H_2O \rightarrow Fe(OH)_3 + 3H^+ \tag{8}$$

$$14Fe^{+3} + FeS_2 + 8H_2O \rightarrow 15Fe^{+2} + 2SO_4^{-2} + 16H^+ \tag{9}$$

In the presence and absence of oxygen, the oxidation of pyrite is possibly accomplished by Fe(III) at pH values less than 2 with the rate of reaction increasing with an increasing Fe(III)/Fe(II) ratio.[14-16] The reaction is felt to proceed in two steps:

$$FeS_2 + 2Fe^{+3} \rightarrow 3Fe^{+2} + 2S^0 \tag{10}$$

$$8H_2O + 2S^0 + 12Fe^{+3} \rightarrow 12Fe^{+2} + 16H^+ + 2SO_4^{-2} \tag{11}$$

Overall:

$$FeS_2 + 8H_2O + 14Fe^{+3} \rightarrow 15Fe^{+2} + 16H^+ + 2SO_4^{-2} \tag{12}$$

The rate of this reaction is remarkably rapid with the rate of disappearance of Fe(III) dependent upon

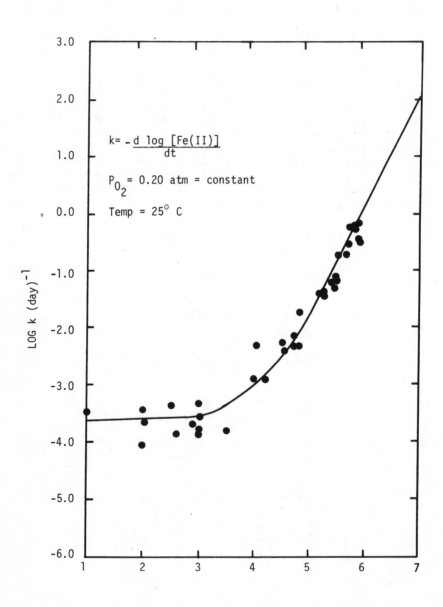

Figure 7.   Oxygenation of ferrous ion as a function of pH.[14]

both the $[FeS_2]$ and $[Fe(III)]$ and independent of $[O_2]$ up to 0.2 atm. When $[Fe(III)] = 10^{-3}$ M and $[FeS_2] = 10^{-2}$ M, the half-time of the reaction is two days. Note that Fe(III) oxidizes FeS_2 much faster than does O_2.

Sulfides and polysulfides may play a role in the oxidation of pyritic ores, especially in the absence of oxygen. It is possible that FeS_2 dissociates directly to Fe^{+2} and S_2^{-2} ions with the subsequent disproportionation of the S_2^{-2} ion in the strongly acidic solution:

$$FeS_2 \rightarrow Fe^{+2} + S_2^{-2} \tag{13}$$

$$S_2^{-2} + 2H^+ \rightarrow H_2S + S^0 \tag{14}$$

Solid marcasite and pyrite are known to exist as discrete Fe^{+2} and S_2^{-2} ions. The reduction of Fe(III) by H_2S in the absence of bacteria is a well-known reaction:

$$4H_2O + 8Fe^{+3} + H_2S \rightarrow 8Fe^{+2} + SO_4^{-2} + 10H^+ \tag{15}$$

A DISCUSSION OF A MOST PROBABLE
MECHANISM FOR FeS_2 OXIDATION IN
THE PRESENCE OF OXYGEN AND BACTERIA

The study of the relationship between the chemically and biologically mediated oxidation reactions in pyritic systems has been the basis of considerable effort in recent years. It has been proposed that in the presence of oxygen, the oxidation reactions accompanying the dissolution of pyrite follow very closely the sequence proposed by Temple and Delchamps.[1] Observations of the rate of oxidation of Fe(II) to Fe(III) by oxygen (Figure 7) and the rate of reduction of Fe(III) by pyrite (Figure 8) in the presence and absence of O_2 show that most prior studies of pyritic oxidations have wrongly considered oxygen as the specific oxidant for pyrite. Oxygen is involved only in the regeneration of Fe(III). An alternate source of Fe(III) is precipitated $Fe(OH)_3$ within the mine itself. In accordance with the experimental observations shown in Figures 7 and 8,

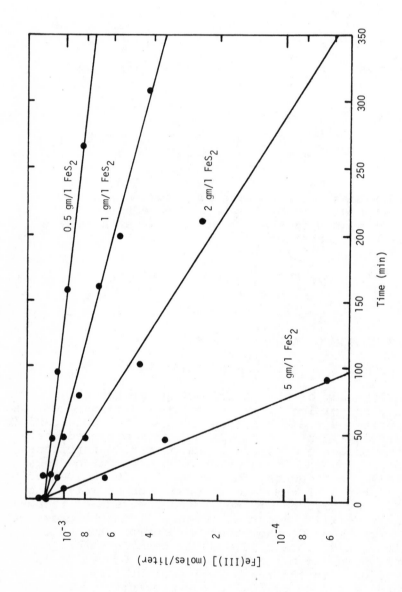

Figure 8. Reduction of ferric iron by pyrite (200-250 mesh) at pH = 1.0. [Fe(III)]$_o$ = 1.3 x 10^{-3} M.[14]

the following model most probably describes the oxidation of pyrite in the presence of oxygen and the appropriate bacterial populations:

1. Chemically, pyrite is converted first to ferrous sulfate

$$FeS_2 + H_2O + \frac{7}{2} O_2 \rightarrow Fe^{+2} + 2H^+ + 2SO_4^{-2}. \qquad (16)$$

2. In the acid environment the ensuing ferrous sulfate undergoes almost no chemical oxidation but is oxidized by *T. ferrooxidans*:

$$2Fe^{+2} + \frac{1}{2} O_2 + 2H^+ \xrightarrow{slow} 2\,Fe^{+3} + H_2O. \qquad (17)$$

3. The Fe(III) formed above is then rapidly converted as it is formed according to the following reaction:

$$2Fe^{+3} + FeS_2 \xrightarrow{fast} 3\,Fe^{+2} + 2S^0. \qquad (18)$$

4. The S^0 resulting from the oxidation of pyrite is then oxidized by ferric ion:

$$2S^0 + 12Fe^{+3} + 8H_2O \rightarrow 12Fe^{+2} + 2SO_4^{-2} + 16H^+ \qquad (19)$$

or by oxygen, in which case the S produced in Step 3 is converted to sulfate by *T. thiooxidans*:

$$2S^0 + 3O_2 + 2H_2O \xrightarrow{T.\ thiooxidans} 2SO_4^{-2} + 4H^+ \qquad (20)$$

The Fe^{+2} produced is then subjected to further microbial action by *T. ferrooxidans* by means of Step 2. A cycle is thus established in which Fe(III) formed by microbial action reacts with pyrite:

$$FeS_2 + 14\,Fe^{+3} + 8H_2O \rightarrow 15Fe^{+2} + 2SO_4^{-2} + 16H^+ \qquad (21)$$

This mechanism is particularly appealing in that the microbially mediated reactions included are known to occur in the presence of the respective bacteria under laboratory conditions. In the absence of bacteria the reactions proceed as indicated except that they occur by strictly chemical means.

Observations of the rate of oxidation of Fe(II) to Fe(III) in the presence of microorganisms isolated from mine drainage waters show that at a pH of *ca.* 3 this reaction proceeds much more rapidly than in laboratory studies conducted with sterile medium (Figure 9). The reaction is also zero order in Fe(II), indicating microbial catalysis of the reaction.

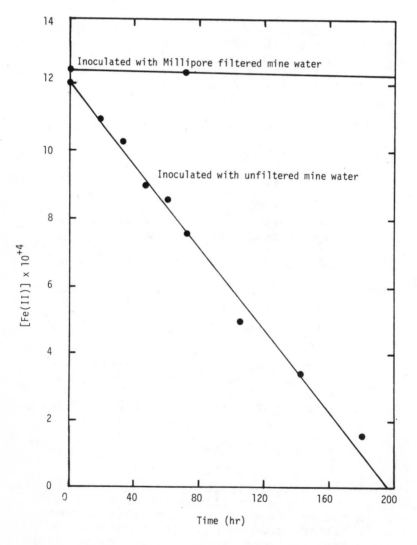

Figure 9. *Oxidation of Fe(II) solutions inoculated with filtered and unfiltered mine water.*[14]

On the basis of the earlier laboratory studies presented, it is also likely that *T. thiooxidans* does not participate in the oxidation of sulfide. The fact that mixed cultures of the two species oxidized FeS_2 no more rapidly than in cultures containing *T. ferrooxidans* alone also supports this mechanism since the oxidation of S^0 by *T. thiooxidans* does not affect the rate of reaction.

The first reaction in the sequence is the oxidation of pyrite. The sulfur species formed in this step may or may not be SO_4^{-2} as indicated, because it is very possible that the first step in the reaction sequence is simply the dissociation of FeS_2 into Fe^{+2} and S_2^{-2} with the eventual production of elemental sulfur as suggested in the previous section.

Since the first step can serve as an initiation reaction in beginning the Fe(II)-Fe(III) cycle established through Steps 2 to 4, the elimination of oxygen is important only in terms of regenerating Fe(III). This is the most interesting feature of the entire reaction sequence. Air-saturated water, once isolated from the atmosphere, contains enough O_2 to form 500 ppm Fe^{+3} by means of Step 2 at 0° C. The fact that the oxidation of Fe^{+2} in sulfate medium is stoichiometric with oxygen uptake and is mediated by *T. ferrooxidans* also lends support for the occurrence of this step. The oxidation reaction is also observed to cease when O_2 is excluded from the cell suspension[11] as is the case in cell suspensions containing Fe(II) only.

The oxidation of Fe^{+2} in Step 2 is the rate-determining step in the overall sequence since the direct oxidation of FeS_2 by Fe^{+3} in Step 4 is a very rapid reaction. The presence of aerobic bacteria markedly increases the rate of pyrite oxidation by enhancing the oxidation of Fe^{+2}. The entire reaction sequence would therefore stop when entrapped O_2 is consumed.

As the anoxic water leaves the mine and is mixed with carbonate-bearing waters, Step 3 is no longer important and the reaction sequence is modified by the precipitation of $Fe(OH)_3$ at the higher pH values:

$$FeS_2 + H_2O + \frac{7}{2} O_2 \rightarrow Fe^{+2} + 2SO_4^{-2} + 2H^+ \qquad (22)$$

$$Fe^{+2} + \frac{1}{4} O_2 + \frac{5}{2} H_2O \xrightarrow{\textit{T. ferrooxidans}} Fe(OH)_3 + 2H^+ \qquad (23)$$

Several aspects of acid formation are not explained by the Temple and Delchamps-type mechanism. First, if Step 1 of the reaction sequence serves merely to initiate the reaction, reactivity differences between pyrite and marcasite should have little effect on the rate of acid formation. Field observations have shown that the rate of generation increases where marcasite is present. A smaller particle size also enhances the rate of reaction. Second, intermediate sulfur species formed during the reaction sequence have not been experimentally established. This might be particularly enlightening when one considers the enzymatic reactions involved in the oxidation of reduced sulfur species. The reaction sequence also does not explain the accumulation of Fe(III) species often found in deep mine water void of oxygen (Table X), although this discrepancy can be explained in terms of mass transport.

REFERENCES

1. Temple, K. L. and E. W. Delchamps. "Autotrophic Bacteria and the Formation of Acid in Bituminous Coal Mines," *Appl. Microbiol., 1,* 255 (1953).
2. Barnes, H. L. and S. B. Romberger. "Chemical Aspects of Acid Mine Drainage," *JWPCF, 40,* 371 (1968).
3. Lees, H. *Biochemistry of Autotrophic Bacteria.* (London: Butterworths, 1955).
4. Parker, C. D. and J. Prisk. "The Oxidation of Inorganic Compounds of Sulfur by Various Sulfur Bacteria," *J. Gen. Microbiol., 8,* 344 (1953).
5. Beck, J. V. "A Ferrous Iron Oxidizing Bacterium," *J. Bacteriol., 79,* 502 (1960).
6. Smith, E. E. and K. S. Shumate. *The Sulfide to Sulfate Reaction Mechanism,* Federal Water Pollution Control Research Series, Washington, D.C. (February, 1970), pp. 1-2.
7. Colmer, A. R. and M. E. Hinkle. "The Role of Microorganisms in Acid Mine Drainage," *Science, 106,* 253 (1947).
8. Colmer, A. R., K. L. Temple, and M. E. Hinkle. "An Iron Oxidizing Bacterium From the Acid Drainage of Bituminous Coal Mines," *J. Bacteriol., 59,* 317 (1950).
9. Vishniac, W. and M. Santer. "The Thiobacilli," *Bacteriol. Rev., 21,* 195 (1957).
10. Leathen, W. W., S. A. Braley, and L. S. McIntyre. "The Role of Bacteria in the Formation of Acid from Certain Sulfuritic Constituents Associated with Bituminous Coal. I. *Thiobacillus Thiooxidans,*" *Appl. Microbiol., 1,* 61 (1953).

11. Brynner, L. C., I. V. Beck, D. Davis, and D. Wilson. "Microorganisms in Leaching Sulfide Minerals," *Ind. Eng. Chem., 46,* 2587 (1954).

12. Waksman, S. A. and I. S. Joffe. "Microorganisms Concerned with the Oxidation of Sulfur in Soil II," *J. Bacteriol., 7,* 2 (1922).

13. Sokolova, G. A. and G. Karavaiko. II, *Physiology and Geochemical Activity of Thiobacilli,* Translated from Russian (Springfield, Va.: Clearing House for Federal Scientific and Technical Information, 1968).

14. *Oxygenation of Ferrous Iron.* Harvard University, Federal Water Quality Administration Water Pollution Control Research Series 14010-06169 (June, 1970).

15. Garrels, R. H. and M. W. Thompson. "Oxidation of Pyrite by Iron Sulfate Solutions," *Amer. J. Sci., 258-A,* 57 (1960).

16. McKay, O. R. and J. Halpern. "A Kinetic Study of the Oxidation of Pyrite in Aqueous Suspension," *Trans. Met. Soc., AIME, 212,* 301 (1958).

17. Peck, H. D. and E. Fisher. "Comparative Metabolism of Sulfur Compounds in Microorganisms," *Bacteriol. Rev., 26,* 1 (1962).

8. SPECIFIC ENVIRONMENTAL PROBLEMS ASSOCIATED WITH THE PROCESSING OF MINERALS

Don H. Baker, Jr.
Director New Mexico Bureau of Mines
and Mineral Resources
New Mexico Institute of Mining and Technology
Socorro, New Mexico

Roshan B. Bhappu
Vice President Mountain States
Research and Development
Tucson, Arizona

ABSTRACT

This paper considers specific environmental problems associated with the processing of minerals. The topics covered include the development of special attributes of mineral processing engineers as pollution experts; a discussion on the environmental problems associated with the use of common and specific flotation reagents; an examination of the effect of hazardous metallic ions in plant effluents; and a plea for the incorporation of environmental awareness and training as integral parts of mineral engineering curricula.

INTRODUCTION

In his persistent efforts to satisfy his insatiable demand for mineral resources, man has often exploited them without regard to consequence. It would be fruitless to deny that the mineral industry by its very nature contributes to certain pollution problems. Open pit mining, underground operations,

hydraulic mining, oil and gas production, metal
smelting, fuel refining and all other extractive
efforts disturb our environment to some degree.
Some damage to the environment is inevitable even
with the best extraction procedure and land restoration methods. The mined ores are moved and handled
several times before they are processed, thus
creating dust, noise, and other forms of air pollution. Mining wastes and drainage usually contribute
to stream pollution and create a solid waste disposal
problem.

It is for these reasons that the mineral industry
is identified by a large segment of the uninformed
public as a major contributor to the overall pollution problem. This is far from the truth. In the
first place, only a small percentage (about 0.14
per cent) of the total land in the United States has
been adversely affected by mining. Moreover, practically no public attention has been drawn to the
extensive as well as expensive efforts made by the
mineral industry to minimize the environmental
impact of its operations. Many extractive companies
take great pride in the responsible manner in which
they have conducted their business. Their efforts
in the environmental control field have gained
national recognition and praise from government
agencies as well as the public.

As members of a progressive engineering society
it is our responsibility to achieve maximum conservation of a finite supply of minerals with minimum
contamination of the environment. We must continue
to develop and implement operational practices,
processes and techniques to achieve the above goal
and to assure full consideration of health, safety
and welfare of the public as well as the people
working in our industry.

As mineral processing engineers we must do our
part in creating a safe and pollution-free environment in and around our mills and processing plants.
Moreover, because of our special expertise in understanding pollution situations and in prescribing
economic solutions to such problems, it is imperative
that we provide the needed lead not only in our own
industry but also in other fields of endeavor.

It is not the intention of the authors to cover
in detail the environmental problems associated with
mineral processing.[1-4] Rather, the primary objectives are to develop the special attributes of
mineral processing engineers as pollution experts,
to pinpoint some specific problems encountered in
different phases of processing, to discuss in some

detail the environmental problems associated with the use of common and specific flotation reagents, to examine the effect of hazardous metallic ions in plant effluents and to plead for the incorporation of environmental awareness and training as integral parts of mineral engineering curricula.

MINERAL PROCESS ENGINEERING
AND POLLUTION CONTROL

Since the art and science of metallurgy deals in the separation of valuable material from waste products either in the solid, liquid or gaseous form, it is easy to appreciate that mineral processing engineers have a special competence for understanding and remedying pollution situations. The same basic principles that govern metallurgical processes also control pollution abatement technology, and most of the common processing machinery can be applicable to pollution problems. Classifiers, cyclones, electrostatic precipitators, thickeners, flotation cells or aerators, digesters, filters, extractors, and the like are all mineral processing units that have found specific applications in pollution control schemes.

According to Aplan[5] the essence of mineral process engineering consists of particle technology, applied surface chemistry, and chemical dissolution which are also basic elements of a pollution control program. Furthermore, the technology of solid waste disposal, recovery of secondary metals and recycling all depend on the unit operations and processes that are the basis of separation of metals and minerals from their ores. A typical flowsheet proposed by Sullivan *et al.*[6] for waste disposal as illustrated in Figure 10 clearly reveals that it is comparable to the flowsheet for recovery of minerals from their ores.

For these reasons, it is up to the metallurgists and mineral processing engineers to pave the way for improvements and innovations in the pollution abatement technology and to share their knowledge with environmental engineers in providing the badly needed mineral resources and metals with a minimum degradation of the environment. Also for these reasons, there is no excuse for the absence of abatement systems in mill design and the existence of pollution and environmental problems in processing plant operations.

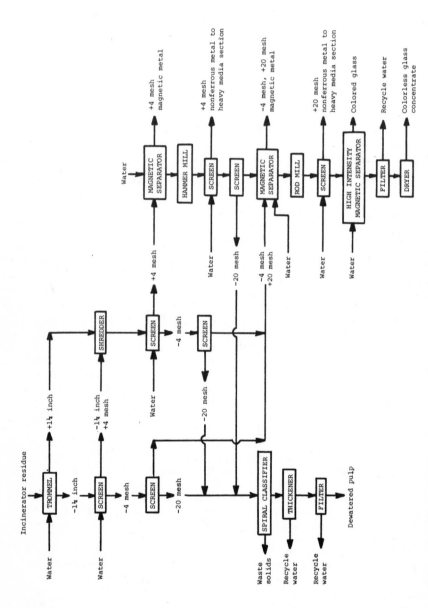

Figure 10.   Physical beneficiation flowsheet for concentrating valuable materials contained in incinerator residues (according to USBM).

ENVIRONMENTAL EFFECTS OF
FLOTATION PROCESSING

Froth flotation is the most widely used technique
for the recovery of valuable minerals from ores. In
this process, the disposal of the solids in the
tailings and the residual reagents and soluble metal
ions and complexes contained in the liquid portion
of the tailings constitute the major pollution
problems.

The task of controlling and abating the extent
of pollution problems from a flotation circuit should
be critically evaluated during laboratory-pilot plant
testing. The mineral processing engineer designing
or evaluating a given flotation operation should
consider critically not only the per cent recovery,
concentrate grade and overall reagent costs, but
also the environmental problems associated with each
flotation scheme.

In the future, it is anticipated that most of
the water control boards and environmental protection
and improvement agencies will demand environmental
position or feasibility reports not only from newly
proposed flotation plants but also from operating
plants.

The major contributors to the pollution problems
in the flotation plant are the rather noxious reagents
used in the process. These include a wide variety
of collectors, frothers, depressants, activators,
and modifiers. Such organic and inorganic reagents
constitute hazardous pollutants in ground or surface
waters due to flotation effluents in the soil,
breakage of tailings dam or pipelines, and accidental
spills.

The pollutants contained in the aqueous portion
of the tailings may be grouped into two major
classes: first, the residual organic and inorganic
reagents added deliberately to the flotation circuit
to recover the valuable minerals, and second the
light and heavy metal ions and complexes contributed
to the aqueous phase by the ore itself due to partial
solubilities of various minerals in water and/or to
their dissolution by some of the flotation reagents.

EFFECT OF FLOTATION REAGENTS

A critical examination of Table XII reveals the
following observations concerning the residual
reagents contained in the aqueous portion of the
tailings:

Table XII

Environmental Considerations for Common Flotation Reagents

Reagent Type & Name	Chemical Composition	Amount Added	Reagents Distribution		Toxic Species
			Solids	Solution	
Collectors					
Xanthate	ROCSSH	<0.1	Mostly complexed	Minor	CS_2
Thiophosphates	(RO)$_2$PSSH	<0.1	Mostly complexed	Minor	$H_2PO_4^-$, HS^-, H_2S
Fatty acids	RCOOH	0.5 to 2.0	Complexed with minerals and ions	Minor	Biodegradable
Sulfonates and Sulfates	RSO_3H RSO_4	0.5 to 2.0	Complexed with minerals and ions	Minor	Biodegradable (except Cyclic)
Amines	RNH_2Ac	0.2 to 2.0	Adsorbed on silicates	Minor	Free amine*
Fuel oils	Saturated	0.1	Selectively adsorbed	Minor	Volatile hydrocarbons
Frothers					
Pine oil	Aromatic alcohols	< 0.1	Carried over in froth	Appreciable	Volatile, carcinogenic
Dow froth	Higher alcohols	< 0.1	Carried over in froth	Appreciable	Volatile, biodegradable
MIBC	Methyl isobutyl-carbinol	< 0.1	Carried over in froth	Appreciable	Volatile, biodegradable
Conditioners					
Sodium sulfides	Na$_2$S, NaHS	1 to 15	Selectively complexed	Major	HS^-, H_2S, $SO_4^=$
Phosphorus Pentasulfide	P_2S_5	1 to 4	Selectively complexed	Major	HS^-, H_2S
Sodium Cyanide	NaCN	0.005 to 0.1	Mostly complexed	Minor	CN^-
Sodium Silicate	Na$_2$SiO$_3$	0.2 to 1.0	Mostly complexed	Appreciable	H_4SiO_4
Milk of Lime	Ca(OH)$_2$	1 to 4	Mostly complexed	Appreciable	OH^-
Separan	Polyacrylamid	<0.01	Mostly adsorbed	Minor	Biodegradable

*Unknown toxicity.

1. The amounts of flotation reagents used in
   the process are very small and kept at a
   minimum (starvation levels) not only because
   of economics but also to avoid the detrimental effects of excess reagents on the
   flotation results and water quality of the
   plant effluents.
2. Most of the added reagents are more or less
   completely consumed during the flotation
   process due to adsorption onto the mineral
   surfaces or complexing with appropriate
   ions in solution.
3. The trace amounts of some organic reagents,
   such as hydrocarbons, alcohols, and esters,
   remaining in the tailings water evaporate,
   decompose or biodegrade.
4. In general, inorganic reagents contribute
   either active cations or anions that react
   at the mineral surfaces or in the aqueous
   phase, leaving one or the other in the
   tailings liquid. Since most of these cations
   or anions are always present in streams and
   natural ground waters, they do not act as
   pollutants in the small amounts added to the
   flotation systems.

From these observations it is evident that the
limited amounts of several reagents added to the
flotation circuits are almost completely consumed
or complexed in the process or are eliminated or
decomposed during the impounding of the tailings.
Therefore, in general, they do not contaminate the
decant water discharged into the stream, river or
lake, nor do they pollute the ground water due to
seepage.

However, some toxic reagents if used indis-
criminately or carelessly in the flotation systems
can constitute a major water pollution hazard and
even a fatal pollutant under plant operating conditions. Cyanide and hydrogen sulfide are two such
reagents that need a more detailed scrutiny. These
reagents are widely used in sulfide flotation systems, especially in the selective flotation of
molybdenite from copper concentrates. Because of
their extremely toxic nature to man, animals and
aquatic life, they have been under investigation
by the state and federal environmental protection
agencies.

Davis[7] has thoroughly covered the topic of control of water pollutants from flotation plants.

The common methods proposed for the treatment of flotation plant effluents prior to their discharge in streams, rivers, lakes or ground water are chemical alteration and precipitation, coagulation and sedimentation, activated charcoal adsorption, biological treatment, filtration, and foaming. In addition to these, the processing engineers should consider some of the newly developed methods such as ion exchange, reverse osmosis and electrodialysis.

The Cyanide Problem

Wright and Bhappu[8] have discussed the environmental problem associated with the use of cyanide in flotation circuits. The primary reason for adding sodium cyanide to the molybdenite flotation circuit is to depress pyrite and other gangue sulfide minerals in the ore so that they will not float along with the molybdenite. This depressing action of cyanide is due primarily to the formation of extremely stable water-wettable complexes with heavy-metal cations in two stages: first, precipitates of relatively insoluble cyanide salts form, and second they dissolve in excess cyanide to form cyanide complexes.

For example, the cyanide ion reacts with the ferrous ion in aqueous solution to form the complex ferrocyanide ion $[Fe(CN)_6]^{-4}$. A similar complex forms with ferrous ions present on the pyrite surface, thus depressing pyrite in the selective flotation of molybdenite. Corresponding cyanide complexes also form on the copper and zinc sulfides if present in the ore.

In flotation systems, the effect of alkali cyanide additions to depress gangue sulfides depends to a large extent on the pH of the system. Sodium cyanide hydrolyzes in aqueous solutions to form free alkali and hydrogen cyanide, relatively insoluble in water:

$$NaCN + H_2O \rightleftharpoons HCN + NaOH \qquad (1)$$

According to this equation, the presence of free alkali in solution shifts the hydrolysis to the left. Since the concentration of CN^- determines its availability for depression of gangue sulfides, a higher pH favors its depression effect (see Table XIII).

Besides complexing with heavy-metal cations on the surfaces of the minerals as well as in the

Table XIII

Influence of pH on Relative Abundance of
CN^- Ions and Dissolved HCN[9]

pH	Millimole/millimole Total Alkali Cyanide	
	As CN^-	As HCN
6.0	0.0005*	0.9995
6.5	0.0015	0.998
7.0	0.0047	0.995
7.5	0.0146	0.985
8.0	0.045	0.955
8.5	0.129	0.871
9.0	0.320	0.680
9.5	0.597	0.403
10.0	0.825	0.175
11.0	0.979	0.021
12.0	0.999	0.001
13.0	1.000	0.000

*To obtain the CN^- in milligrams per milligram of sodium
cyanide, it is sufficient to multiply the numbers in this
column by 0.531.

solution, cyanide also has a tendency to dissolve
sulfide minerals of copper, zinc, and iron. The end
products of these reactions are the corresponding
metal cyanides, thiocyanate, and sulfate. For
example, pyrite reacts with cyanide thus:

$$FeS_2 + 7NaCN + 2O_2 = Na_4Fe(CN)_6 + Na_2SO_4 + NaCNS \qquad (2)$$

From this discussion, one can see that the
cyanide added to the flotation system should be used
up in complexing with metal cations on the mineral
surfaces and in solution or in forming thiocyanate.
Also, because of the presence of appreciable amounts
of gangue sulfides as well as heavy-metal cations in
the flotation plant feed and tailings, no free CN^-
should remain in tailings water. The only cyanide
products that could be present in the tailings water
would be complexed heavy-metal cyanides and
thiocyanate.

Table XIV gives average analyses for free cyanide, thiocyanate, ferrocyanide, sulfide, phosphorus, and metal ions present in the plant effluents from several flotation plants in which molybdenite is selectively floated from copper concentrates with the use of sulfides and cyanides. The results clearly indicate that no free cyanide remains in any of the plant effluents, except a very small amount of free cyanide in the reclaimed water that is recirculated in the plant.

Table XIV

Average Analyses of Effluents from
Selective Sulfide Flotation Plants

	Plant Tailing	*Reclaimed Water*	*Tailing Discharge*
	(all values in ppm)		
Cu	0.01	0.01	< 0.01
Zn	0.02	0.05	0.03
Pb	0.07	0.10	< 0.01
Fe	0.05	0.20	< 0.05
Mn	0.50	0.90	0.10
Ca	71.0	65.0	31.0
Mg	17.5	16.0	6.0
Mo	3.0	3.8	0.0
Cn	< 0.005	< 0.005	N.D.
SCN	0.10	0.12	N.D.
Fe(CN)	N.D.	N.D.	N.D.
$S^=$	N.D.	N.D.	N.D.
P	N.D.	< 0.01	< 0.01

N.D. - none detected

Therefore, the limited amount of cyanide added to the flotation systems does not constitute a hazardous condition either at the mill or in the plant effluent. The only way in which free cyanide can remain in the tailings effluent is if it is added in excess, a condition that should never be allowed to exist in an efficient and safe flotation plant.

In recent years, the removal of residual cyanide from waste effluents has received much attention. Most of the published literature is concerned with the treatment of spent solutions from electroplating plants.[10] Methods for cyanide removal include oxidation processes (with Cl_2, $NaOCl$, O_3, H_2O_2, and electrolysis), ion exchange formation of insoluble cyanide complexes acidification-aeration process, and biological decomposition.

Hydrogen Sulfide Problem

Alkali sulfides have been used extensively in flotation systems both as activators (for oxide or tarnished sulfide minerals) and depressants (for depressing gangue sulfides in selective flotation of particular sulfide minerals). These systems have been critically examined by Cox *et al.*,[11] Gaudin[9] and Dudenkov *et al.*[12] Theoretical work suggests that the actual depressant is the hydrosulfide ion, HS^-, rather than sodium sulfide or hydrogen sulfide.

An examination of Table XV reveals that the availability of HS^- ions increases with increasing pH up to 9.0 after which the concentration of the ions remains more or less constant. Therefore in order to obtain maximum benefit of the reagent, the flotation should be carried out at higher pH values.

From environmental and safety considerations, the major factor to watch for is the liberation of H_2S in the flotation system. Although the solubility of H_2S in water is relatively high (0.12 m/l at one atmosphere), under the existing flotation conditions there are good possibilities for evolution of H_2S, thus causing a hazardous operating condition. It is fortunate that sufficient dilution of H_2S with fresh air takes place in most of the operating plant to keep its concentration below the threshold level. Nevertheless, the problem does exist and efforts should be made to keep down the H_2S concentration through suitable operating controls and proper ventilation.

The toxicity of H_2S is greatly underrated. In sewage disposal, 1 ppm of sulfide under certain conditions (mostly acidic) causes a most lethal atmosphere.[13] Lethal concentrations for H_2S are similar to those for hydrocyanic acid. However, the latter is more soluble in water and thus somewhat less hazardous.

Table XV

Relative Abundance of H_2S, HS^- and $S^=$ as a Function of pH[9]

For every millimole $Na_2S \cdot 9H_2O$ added per liter, there is

pH	Dissolved H_2S (millimole/liter)	Dissolved HS^- (millimole/liter)	Dissolved $S^=$ (millimole/liter)
4.0	1.0	0.0010	2×10^{-14}
4.5	1.0	0.0030	2×10^{-13}
5.0	0.99	0.010	2×10^{-12}
5.5	0.97	0.030	2×10^{-11}
6.0	0.91	0.09	2×10^{-10}
6.5	0.75	0.25	1.6×10^{-9}
7.0	0.50	0.50	0.5×10^{-8}
7.5	0.25	0.75	6.5×10^{-8}
8.0	0.09	0.91	2×10^{-7}
8.5	0.030	0.97	7×10^{-7}
9.0	0.010	1.00	0.000002
10.0	0.0010	1.00	0.00002
11.0	0.00010	1.00	0.0002
12.0	0.000010	1.00	0.002
13.0	0.000001	0.98	0.02
14.0	0.0000001	0.80	0.20

Since the concentration of H_2S is dependent on the pH, the safest as well as the most effective method in the plant is to try to maintain a high pH in the flotation circuit with the use of caustic. This is especially true for those circuits using Nokes' reagent containing phosphorus pentasulfide (P_2S_5). The most critical phase of the use of this reagent is in the preparation of Nokes' solution for plant use. It is imperative that ultimate precautions be taken in the mixing step to make sure that all the P_2S_5 has reacted with caustic and that the solution is stabilized. If equilibrium is not reached, decomposition of the reaction product can occur, giving off H_2S.

The optimum conditions can be met by digesting P_2S_5 slowly in at least a 10% excess caustic over the stoichiometric requirement. Since the reaction is exothermic and since the decomposition of the reaction product is enhanced at higher temperatures, precautions must be taken to keep the temperature down, preferably below 120° F. Provisions should also be made to have available at the mixing tank a source of caustic so that, if needed, it can be added to the mixture to prevent its decomposition. Additional safety devices such as alarm systems based on H_2S detection should also be installed at strategic locations and operators should be trained to avert a possible disaster.

Hazardous Metallic Ions in Plant Effluents

As pointed out earlier the mill tailing water contains several light and heavy metal ions and complexes that are normally contributed to the system due to natural solubility of valuable and gangue minerals contained in the ore. Some of these dissolved metals such As, Cr, Hg, Pb and Se constitute serious pollution problems and are considered "hazardous substances" by the environmental protection agencies.

In order to prevent the degradation of the environment and injury to man, animals, and aquatic life, the government agencies have adopted "regulations" concerning the discharge of hazardous substances. Typical regulations adopted by the New Mexico Water Quality Control Commission on August 27, 1971, are as follows:

Regulation Number 6 - Discharge of Hazardous Substances

A. No person shall cause or allow effluent from a new process or plant to discharge directly or indirectly into water as defined by the Water Quality Act unless the effluent, as indicated by any single sample, conforms to the following:

Arsenic	0.05 mg/l or less
Barium	1.0 mg/l or less
Boron	0.75 mg/l or less
Cadmium	0.01 mg/l or less
Chromium (Total)	0.01 mg/l or less
Copper*	0.05 mg/l or less
Lead	0.05 mg/l or less

Manganese	0.1 mg/l or less
Mercury	0.001 mg/l or less
Molybdenum	0.01 mg/l or less
Nickel	0.1 mg/l or less
Selenium	0.01 mg/l or less
Silver	0.05 mg/l or less
Zinc*	0.1 mg/l or less

*Provided that effluents from a community sewerage system may contain 0.1 mg/l copper and 0.5 mg/l zinc.

B. Subsection A does not apply to constituents in the water diverted. Samples shall be examined in accordance with the latest editions of Standard Methods for the Examination of Water and Wastewater published by the American Public Health Association, or Methods for Chemical Analysis of Water and Wastes published by the Environmental Protection Agency, where applicable.

Regulation Number 4 - Effluent Quality

A. Except as otherwise provided in this regulation, no person shall cause or allow effluent to discharge directly or indirectly into water as defined by the Water Quality Act unless the effluent:

1. as indicated by any daily composite sample, conforms to the following:

Biochemical Oxygen Demand (BOD)	less than 30 mg/l
Chemical Oxygen Demand (COD)	less than 125 mg/l
Settleable Solids	less than 0.5 ml/l

2. as indicated by single sample, conforms to the following:

Biochemical Oxygen Demand (BOD)	less than 160 mg/l
Chemical Oxygen Demand (COD)	less than 250 mg/l
Settleable solids	less than 1.0 ml/l
Fecal coliforms	less than 500 organisms/ 100 ml
pH	between 6.7 and 8.6

Upon application, the Director of the Environmental Improvement Agency may eliminate the pH requirement for any effluent source that the Director determines does not unreasonably degrade the water into which the effluent is discharged.

As mineral process engineers we can appreciate the primary reasons for the very low concentration of metallic ions in tailing effluents. First, the solubility of many naturally occurring minerals (sulfides, oxides and silicates) is limited. Second, most of the dissolved metallic ions tend to hydrolyze at higher pH values normally used in the flotation systems and since the solubilities of most of the metal hydroxides are limited (see Table XVI), the ionic concentrations are correspondingly low (well below the specified regulations). Moreover, some of the metal ions such as Mo, W and V complex with calcium (available through lime addition to the flotation circuit) to form very insoluble compounds such as $CaMoO_4$ and $CaWO_4$. Finally, many light and heavy metallic ions are complexed with flotation reagents (collectors, depressants and modifiers) to form insoluble complexes. The end result of all these chemical reactions in the flotation system is the depletion of hazardous ions in the tailing effluents.

In the event that concentrations of hazardous ions in the plant effluent are above specified limits, then water must be treated to comply with the regulations. Methods commonly used in the tertiary treatment of municipal waters and abatement of pollutants in industrial effluents can be readily applicable to the treatment of mill effluents. These include neutralization, chemical alteration and precipitation, charcoal adsorption, ion exchange, biological treatment and foam flotation.

The above discussion clearly indicates that with proper understanding of the mineral-chemistry, surface-chemistry, ionic equilibria and the flotation mechanisms coupled with the abatement technology, it should be possible to economically remove hazardous substances from plant effluents. Certainly, the technology and the proper incentives are available to the mineral process engineers to operate economically without polluting the environment.

Table XVI

*Solubility of Selected Minerals, Precipitates and
Organo-Metallic Complexes*

Minerals or Compounds	Solubility in Water (ppm)	
	Cations	Anions
Cu_2S	2.2×10^{-14}	5.4×10^{-12}
CuS	1.9×10^{-7}	9.6×10^{-8}
PbS	5.8×10^{-4}	9.0×10^{-10}
HgS	1.26×10^{-21}	2.0×10^{-23}
NiS	3.2×10^{-5}	1.7×10^{-5}
ZnS	2.0×10^{-6}	1.0×10^{-6}
$PbWO_4$	12	5.7
$PbSO_4$	0.8×10^3	0.38×10^3
$CaCO_3$	3.7×10^3	5.6×10^3
CaF_2	8.6×10^{-3}	8.2×10^{-3}
$Ca_3(PO_4)_2$	8.4×10^{-4}	1.26×10^{-3}
$CaSO_4$	0.52×10^3	1.26×10^3

	Hydrolysis Product (Ksp)	pH of Precipitation
$Fe(OH)_2$	1×10^{-15}	7.0
$Fe(OH)_3$	1×10^{-38}	3.0
$Al(OH)_3$	1×10^{-32}	4.2
$Cu(OH)_2$	1×10^{-20}	5.5
$Mg(OH)_2$	6×10^{-12}	10.5
$Ca(OH)_2$	8×10^{-6}	11.5

	Solubility Products (Ksp)
Zn - ethyl xanthate	5.3×10^{-9}
Pb - ethyl xanthate	6.7×10^{-16}
$Ca(Ol)_2$	3.9×10^{-13}
$Mg(Ol)_2$	1.6×10^{-11}
$Ba(Ol)_3$	1.3×10^{-12}
$Fe(Ol)_2$	3.9×10^{-13}
$Mg(RSO_3)_2$	10^{-8} to 10^{-11}
$Ca(RSO_3)_2$	10^{-8} to 10^{-14}
$Fe(RSO_3)_2$	10^{-18} to 10^{-21}
$Al(RSO_3)_2$	10^{-16} to 10^{-18}

NEED FOR ENVIRONMENTAL AWARENESS AND TRAINING

It is obvious that the mineral process engineer of today and the future has to play a dual role, that of a metallurgist as well as environmentalist. Traditionally, he is trained in the basic sciences and engineering common to both fields. What is needed is a keen awareness and concern to fulfill the dual responsibility that our mineral-based society has imposed on him.

This awareness and technical competence can only be developed through a realistic and well-balanced engineering curriculum. It is not necessary that the undergraduate engineer be required to take a specific course or courses in environmental science and engineering. However, it is essential that the student be made aware of the similarities in the two fields throughout all the engineering courses. With proper orientation and a minimum of deviation an instructor can do an adequate job of training the mineral process engineer so that he will be in a position to solve the environmental problems facing him in his professional career.

REFERENCES

1. "Surface Mining and Our Environment," United States Bureau of Mines, Washington, D.C. (1967).
2. McNay, L. M. "Mining and Milling Waste Disposal Problems— Where are We Today?" Proceedings of Second Minerals Waste Utilization Symposium, U.S.B.M.-IIT Research Institute (March 1970).
3. Dean, K. C., *et al.* "Chemical and Vegetative Stabilization of a Nevada Copper Porphyry Mill Tailings," Report of Investigation 7261, USBM.
4. "Coal Preparation Today," *Coal Age, 73(7),* 227 (1968).
5. Aplan, F. "Mineral Processing Engineers are Experts in Pollution Control Problems," *Mining Engineering, 22(5),* 50 (1970).
6. Sullivan, P. and M. Stanczyke. "Economics of Recycling Metals and Minerals from Urban Refuse," Technical Progress Report #33, USBM. (1971).
7. Davis, F. T. "The Control of Water Pollution from Flotation Plants," AIME Pacific Southwest Mineral Industry Conference (1970).
8. Wright, I. and R. B. Bhappu. "Waste Problems Relative to Mining and Milling of Molybdenum," Purdue University Eng. Bull. Ext. Ser. #129 (Part II) (1967), pp. 579-592.

9. Gaudin, A. M. *Flotation*, 2nd ed. (New York: McGraw-Hill Book Co., 1957).

10. Barnard, P., *et al.* "Recovery of Metallurgical Values from Industrial Wastes," Proceedings of Second Minerals Waste Utilization Symposium, U.S.B.M.-IIT Research Institute, (March, 1970).

11. Cox, A. B. and E. W. Wark. "Principles of Flotation IV-An Experimental Study of Influence of Sodium Sulfide, Alkalies, and Copper Sulfate on Effect of Xanthates at Mineral Surfaces," *Transactions AIME, 134,* 7 (1939).

12. Dudenkov, S. V., *et al. Theory and Practice of Flotation Reagents Use.* edition Nedva (Moscow, 1969).

13. Bowlus, F. and R. Pomeroy. "Stay Alive in That Sewer," *The American City* (1959).

9. THE DETERIORATION OF WATER RESOURCES IN MEXICO AS A RESULT OF INDUSTRIAL WASTE DISPOSAL

Armando P. Baez
Instituto de Geofisica, UNAM
Instituto Nacional de Energia Nuclear
Mexico, D.F.

ABSTRACT

This paper reviews several serious industrial waste pollution incidents that have recently been identified by this researcher. First, the pollution of Coatzacoalcos River due to the excessive discharge of industrial waste effluents without any treatment is considered. The effect of this pollution on the ecology of the region and probable abatement techniques are presented. A second case analyzed is the effect of accelerated waste disposal and eutrophication of the Lerma River and Chapala Lake.

Some suggestions for reducing wastes pollution due to industrial discharge in Mexico are set forth.

INTRODUCTION

Mexico is a country in the process of rapid industrialization. Every year many new industries are created, and those already existing are continuously expanding to satisfy the needs of the exponential increase of the country's population. Most of these industries are responsible for significant air and water pollution problems in the country. Control methods are not in existence in a great majority of them, and industries do not know to what extent they should control their sources of air and water pollution.

To a large extent the pollution problem has been
aggravated by poorly planned industrialization, *i.e.*,
most of the industries are concentrated in only a
few cities, Mexico City, Guadalajara, and Monterrey.
Many industries, located at river estuaries and
basins, discharge their effluents without any form
of treatment into the water. For many years water
resources have been badly polluted. Dramatic cases
of river and lake contamination have been reported
by people who depend upon these water resources for
domestic fishing and agricultural uses.

Let me describe very briefly the actual situation
that prevails in the country as far as water pollution is concerned and to what extent this pollution
has altered and damaged the country's water resources.
As a result of the huge consumption of detergents by
population and industry, eutrophication processes
have been taking place in the most important water
bodies, such as Chapala and Yuriria Lakes, Lerma
and Santiago Rivers, and some dams.

A decrease of some species of fish and other
sensitive water pollution aquatic organisms has been
observed in lakes, lagoons, rivers and estuaries.
Even the disappearance of valuable commercial species
has been reported. As a result it has affected the
economy of many fishing villages.

COATZACOALCOS RIVER

Coatzacoalcos River is one of the most important
fluvial routes in Mexico, as well as one of the
rivers most affected by industrial pollution. There
are many fishing villages along the river banks
whose main income is from the river's resources.
However at present, fish, crab, shrimp, oysters and
other species have either disappeared or migrated
from the estuary, its tributaries and its lagoons.
The increase of fauna mortality in the region has
been paralleled by the industrial development of the
area. Because of lack of pollution control at the
oil refinery, loading docks, and oil tankers, one
can observe a thin film of petroleum on the surface
of the water from the refinery 20 miles upstream to
the delta of the river. This film prevents the occurrence of aeration processes necessary for preserva-
tion of aquatic life in still waters, and only the
strong currents and tides provide mixing processes
to increase oxygen in the river. On the other hand,
oil contributes to depletion of the dissolved oxygen

content of the river water, due to its high BOD demand.

Besides the oil refinery there are many other industries that account for overall load of water pollutants. Sulfur and sulfur compounds, calcium chloride, super phosphates, phenols, vinylchloride, and acetaldehyde are among the most common water contaminants. Significant levels of lead have been found in the water and associated biota near the river's mouth as a consequence of the emissions from the tetraethyl lead factory.

Mercury

In our laboratory we have undertaken with the sponsorship of the Mexican Institute of Nuclear Energy, a research project in the Minatitlán-Coatzacoalcos region aimed at studying the transport of environmental mercury through the food-chain. We have identified two industries that use mercury in their processes. One of these industries is a chloralkali plant, in which mercury is used as a flowing cathode in the mercury-cell process. The other industry produces vinyl chloride. Both of these industries can be considered as potential mercury polluters.

For the last five months we have run a mercury monitoring program. The samples are comprised of water, river bottom mud, aquatic organisms, soil, aquatic and river bank vegetation, and human hair. We have also taken samples from a control site located 20 miles upstream in an area where there is no industry located and with only small fishing villages along the river.

From our preliminary results, we can observe that the Teapa Creek is the most severely contaminated area. The creek discharges into the Coatzacoalcos River where mixing and dilution processes take place, bringing the mercury concentration levels down to very low values.

Although the dilution rate between the Teapa Creek and the Coatzacoalcos River is considerable, we have found mercury concentration factors up to 2,000 and more between water and biota. For example, concentration factors as high as 2,000 have been found in white clam shells, over 4,000 in catfish stomachal content, and over 500 in fish muscles and river vegetation (Table XVII).

Table XVII

Range of Mercury Values Found in Organisms, Water,
and Bottom Muds from the River Estuary and Teapa Creek

Item	Range
Leaves	<0.01 - 10.0 ppm
Flowers	0.07 - 0.27 ppm
Fish Muscle	<0.01 - 36.0 ppm
Fish Stomachal Content	0.22 - 8.1 ppm
Surface Water	<0.01 - 5.0 ppb
Deep Water	<0.01 - 39.0 ppb

LERMA RIVER

The Lerma River is another case of severe water pollution, resulting from the great consumption of detergents by population and industry as well as increasing waste disposal. The sewage from cities nearby as well as many types of industrial effluents are dumped into the river without treatment.

A great part of the river, the dams and lakes that form part of the Lerma River Basin are partially eutrophicated. In some parts we have found amounts of total phosphate as high as 30 mg/l and from 5-10 mg/l of alkyl benzene sulfonate.

Water Pollution Prevention
and Control in Mexico

The Federal Law for the prevention and control of environmental pollution was approved by Congress in 1971 and published in the Federal Register. Rules and regulations for the prevention and control of water pollution were approved by Congress in April of 1973 and published for its enforcement in May of the same year.

These regulations are now in operation, but due to the complex problems that industry is facing we will have to wait some time in order to see if these regulations are in fact operative. On the other hand it will be necessary for industry to be given the necessary technical assistance to attain the regulations. Hopefully the joint efforts of industry and government will positively reduce the water pollution problems that Mexico has today.

ACKNOWLEDGMENT

The author wishes to acknowledge the assistance of Ann Lowe of the Institute of Geophysics for the review of the manuscript.

10. THE APPLICATION OF INFRARED TECHNOLOGY TO ECOLOGICAL PROBLEMS

Lt. Colonel Roland D. Mower, USAF
Wright State University and
Air Force Systems Command (WPAFB)
Dayton, Ohio

Leonard W. Crouch
Air Force Systems Command (WPAFB)
Dayton, Ohio

ABSTRACT

This introduction to thermal infrared (IR) remote sensing techniques stresses the application of infrared technology to ecological problems. The nature of infrared energy and its interaction with environmental parameters such as altitude, time of day, atmospheric attenuation, scene phenomena and sensor performance criteria are discussed. Selected samples of imagery are included to demonstrate the application of infrared sensing techniques to specific atmospheric and waterway pollution problems.

INTRODUCTION

This paper, based upon research conducted by members of the Reconnaissance and Surveillance Division of the Air Force Avionics Laboratory (AFAL/RSE), considers the use of thermal infrared remote sensing techniques for locating and identifying sources of selected atmospheric and waterway pollutants. First, the nature of infrared energy and its interaction with various environmental parameters

101

such as altitude, time of day, atmospheric attenuation, scene phenomena and sensor performance are briefly described. Then, selected samples of imagery are presented and discussed to demonstrate the application of infrared sensing techniques to specific atmospheric and waterway pollution problems.

GENERAL DISCUSSION

Infrared Energy

Infrared radiation is generated by the vibration and rotation of atoms and molecules in any material whose temperature is above absolute (-273°C). As the temperature of a material increases, energy in the form of infrared radiation also increases. Infrared energy falls between the visible light and microwave regions of the electromagnetic spectrum; as a result, it exhibits some characteristics of both visible light and microwave energy (Figure 11).

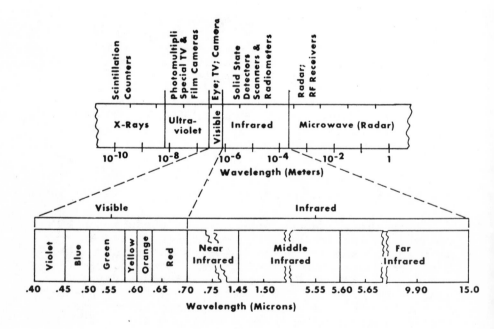

Figure 11. The electromagnetic spectrum.

For example, infrared energy can be optically focused, will propagate through certain materials that are opaque to visible light and yet cannot pass through common optical glass.

The natural process by which the earth receives energy from the sun and reradiates it back into space must be understood and considered by those engaged in remote sensing. During the radiation cycle the earth absorbs the sun's energy at shorter wavelengths, becomes warm, and then reradiates energy in longer wavelength form back to space. More specifically, the energy peak from the sun is at about 0.5 microns (micrometer), whereas reradiation from the earth peaks at approximately 10 microns (this concept is illustrated in Figure 12). Normally man-made heat sources and thermal pollution also radiate energy in the longer infrared wavelength regions.

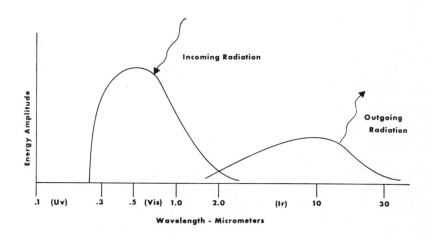

Figure 12. Visible and infrared spectrum.

The absorption, transmission, and spectral distribution of infrared radiation are governed by various physical laws. To understand these parameters one must have some understanding of such concepts as "blackbody" and "emissivity." A blackbody is defined as a perfect absorber of all incident energy with zero reflection. A theoretical blackbody is a perfect radiator of energy and emits the maximum amount of energy possible for its temperature. Emissivity is the ratio of radiance from a material to that of

a perfect radiator or blackbody. The greatest value
of emissivity is unity (1.0), and it occurs only
with a theoretical blackbody. On the other hand, a
body that reflects all incident energy and absorbs
or radiates none has an emissivity of zero. Carbon
black is an example of a material with high emissivity, whereas a highly polished metal surface has
an emissivity value near zero. Most natural materials
fall between these two extremes and are termed gray
bodies.

Several significant factors influence the storage
and transfer of heat within the solid, liquid and
gaseous substances that compose the earth's surface
and atmosphere. Obviously, the relationships between
such factors and the distribution of heat in both
the temporal and spatial context must be understood
if meaningful interpretations of thermal imagery
are to be made. This is especially true if one in-
tends to detect and measure man-made changes to the
natural thermal landscape. For example, the dynamic
nature of the daily and seasonal temperature variations that characterize various geographic and
climatic regions of the world must be recognized
and accounted for.

The spectral distribution of thermal energy is
another extremely important factor that relates
directly to infrared sensing. According to Planck's
law, as the absolute temperature of a body increases,
the peak radiation shifts toward the shorter wavelengths. This concept can be seen when a metal bar
is heated. At first, the bar exhibits its own color
but it slowly turns dull red as the temperature increases. Additional heating causes the color to
change to bright orange and eventually white hot.
Because of unique properties of thermal energy,
infrared is ideally suited for detecting both thermal
and chemical pollution. Infrared detectors are normally designed so that their spectral response
covers selected wavelengths that are dependent upon
the anticipated range of target temperatures.

As infrared radiation passes through the atmosphere various gases act to scatter and attenuate
the energy signal. Such absorption and scattering
are selective in nature and occur in narrow spectral
regions of the infrared band. The principal agents
of attenuation are carbon dioxide and water vapor.
Those portions of the electromagnetic spectrum that
can be transmitted through the atmosphere are termed
windows. Infrared windows occur roughly in the 1.5
to 1.8, 2.0 to 2.5, 3.0 to 4.2, 4.4 to 5.5 and 8.5
to 13.0 micron bands. The latter, 8.5 to 13.0

microns, is the widest window and it corresponds to the range of radiant energy emitted by most surfaces at their normal ambient temperatures.

Infrared Sensors

In order to utilize infrared sensing techniques, one must be able to detect, measure and display emitted energy. Infrared imaging sensors, under military development since about 1950, are generally known by a variety of different names such as thermal imagers, thermal viewers, or IR line scanners. Basically, an infrared imaging sensor consists of several components which include an infrared optical lens assembly, an infrared detector unit, supporting electronics, and a recording device or display (Figure 13).

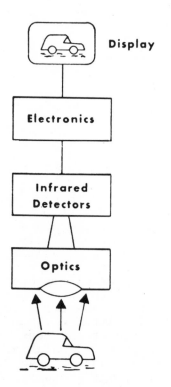

Figure 13. Infrared imaging sensor.

The basic operating principle of IR scanners and mappers is illustrated in Figure 14. These devices utilize a reflective optical system to scan and focus the infrared radiation onto a sensitive detector array that produces an electrical signal. The

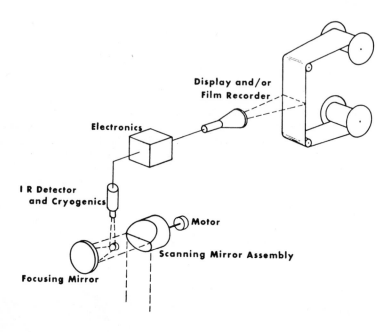

Figure 14.   IR line scan sensor.

resultant signal is amplified and used to modulate a light source that is recorded on conventional photographic film. Detectors using solid state crystals such as indium antimonide or specially doped germanium or tellurium are cryogenically cooled to eliminate ambient effects and increase sensitivity. Detector materials have various spectral responses and are selected to correspond with the atmospheric transmission windows and radiation characteristics of the scene under investigation. Extremely hot targets such as smelters and volcanoes are best sensed with indium antimonide detectors, which are sensitive in the 1.0 to 6.0 micron band. Conversely, doped germanium and trimetal detectors are employed for near ambient temperature targets. A mercury-cadmium-telluride detector element, responsive in the 2.0 to 14.0 micron region, was used for this study.

The supporting electronics system couples the
varying DC signal output of the detector to the
final display. Traditionally, two different approaches
for IR scanner system electronics have been advocated.
One approach amplifies the varying DC signal and records or displays it directly. If calibration
signals are added, the varying DC signal can be
mathematically related to near absolute temperature
of the original surface scene. The second approach
takes the varying DC signal from the detector and
converts it to an AC signal, which is electronically
enhanced for the final display. Both techniques
have certain advantages and disadvantages and are
used when needed. The IR scanner used for this study
was AC coupled and it did not include a calibrated
DC reference. As a result, only relative temperatures
within the scene could be assessed.

An IR radiometer is a device closely related to
IR imaging sensors. The radiometer, however, is used
primarily for quantitatively measuring thermal infrared energy emitted by remote objects. This allows
one to remotely determine the precise effective
temperature of an object and present the value on a
meter rather than in image form. Data acquired by
radiometers can be used to calibrate the output of
imaging sensors. For example, one can use an airborne IR imaging sensor to locate thermal pollution
in a waterway; then an IR radiometer can be used to
determine specific effective temperatures of the
pollution. Another alternative, one which involves
a significant increase in complexity, combines an
IR imager and radiometer into a single unit called
an IR imaging radiometer. A few such devices have
been made and demonstrated.

Infrared Imagery

Infrared radiation is generally collected in a
line scan pattern with the sensor scanning in a
direction perpendicular to the direction of flight.
The forward motion of the aircraft moves the scan
in the direction of flight as shown in Figure 15.
Ground objects that are hot show up as black areas
on an image negative, whereas cooler objects appear
as lighter areas. This tonal characteristic is
reversed when positive prints are made. The tonal
quality or gray level (shade between black and white)
depends primarily upon the temperature and emissivity
of an object in the scene and its background.
Photographs contained in this report are positive

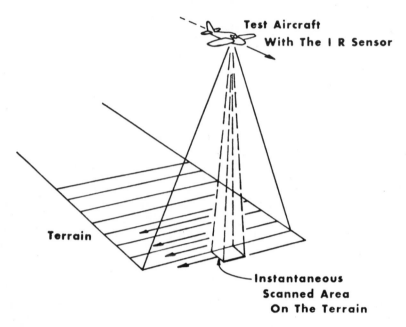

Test Aircraft
With The I R Sensor

Terrain

Instantaneous
Scanned Area
On The Terrain

Figure 15.   Infrared line scanner.

reproductions; warmer objects are depicted by light
tones and cool objects by dark tones.

The USAF thermal mapper used in this project was
AC coupled and measured only the radiation differentials in the ground scene against some drifting
average background tonal density. No absolute correlation between tonal density and target temperatures
can be achieved using this type of infrared device.
Although DC coupled systems are capable of recording
quantitative surface temperature data, they enjoy
very limited use because of their complexity and
cost. An example of a calibrated thermograph
(thermogram) where image tonal scale is referenced
to a calibrated standard is shown in Figure 30.

An IR imaging system samples the scene in discrete lines as shown in Figure 15. When an object
is covered by the width of a single scan line it
will be imaged (in line of flight) as a single dot
the size of the scan line. Larger objects will have
more scan lines placed on them. As a rule, the more
scan lines that cross an object, the more easily an
object is recognized. Indeed, experimental data has
shown that unobscured objects can be "detected" with

four scans, "recognized" with ten scans and "identified" with twenty scans. The number of scan lines
indicated above provides a useful rule of thumb for
flight planning purposes because they help determine
the flight altitude that should be selected based
upon sensor and scene constraints.

MISSION PLANNING CONSIDERATIONS

General

   Flight planning and execution parameters are
very critical for infrared remote sensing. The
spectral region, altitude, time of day, equipment
settings, and prevailing meteorological conditions
can greatly influence the results. Erroneous conclusions can be drawn from imagery acquired when
improper mission planning is used, or when carefully
controlled data collection techniques are not
employed.

Spectral Region

   As previously mentioned, the spectral region
chosen for this study included the 2.0 to 13.0 micron
band, which covers potential target temperatures
ranging from ambient to +800°C. When local hot spot
temperatures, such as those created by small fires,
are the items of interest, the 2.0 to 6.0 micron
region is most useful. However, when target temperatures are close to local ambient temperatures, as
in the case of waterway pollution, the 8.5 to 13.0
micron region is optimum.

Altitude

   The altitude at which an infrared sensor is
flown governs the lateral ground coverage, scale
factor, and ground resolution of the imagery. Higher
altitudes yield wider lateral coverage but at the
expense of scale and resolution. One geographic
region included in this study contained a winding
river. In this particular case a compromise altitude was chosen to provide sufficient lateral
coverage for most of the river bends and at the
same time retain adequate spatial and thermal detail

to identify thermal effluents (outfalls). Additional
flight passes are frequently required in areas where
broad coverage is needed. For example, six separate
flight lines were required to cover the meandering
river network around Pittsburgh, Pennsylvania.

Time-of-Day

Incoming solar radiation significantly affects
the quality of imagery acquired by infrared sensors.
Generally daylight hours are the least favorable
time for conducting remote sensing of many areas of
interest including waterways. The rather poor performance of IR sensors at certain times can be
attributed to several factors. Perhaps most important is the dynamic nature of surface temperatures
that characterize various objects found in the
scene and their interaction with the lower atmosphere. As a result, localized surface winds frequently produce short-term temperature variations
that may exceed the dynamic range of the sensor.
This problem can also occur at night under some
atmospheric conditions.

During daylight hours, reflected solar energy
in the near infrared region (less than 3.7 microns)
also contributes to the scene's apparent temperature.
The introduction of reflected energy to the scene
is not desirable because thermal temperature differentials are determined solely from emitted energy.
The presence of reflected energy (although low in
magnitude) is detrimental because it introduces
unrelated spurious signals to the sensor and the
resultant imagery. This factor can be eliminated
by the use of a short wavelength rejection filter.

A third factor that warrants discussion is the
distribution of temperature on the earth's surface
that results from the specific heat or different
heating rates of the various materials that compose
the scene. This phenomenon can cause thermal
crossovers, a condition that exists during the
heating and cooling cycles of different materials
when they produce the same level of radiant energy.
Crossovers occur most frequently either shortly
after sunrise or after sunset. It should be noted
that in the case of adjacent land and water bodies,
the temperature of the water body changes much
slower than does the temperature of land surface.

Despite the disadvantages of daytime infrared
data collection, successful operation can be

accomplished by proper equipment design and mission planning. For example, filters can be employed to reduce reflected energy in the near infrared region, flights can be scheduled after the early morning hours to avoid most thermal crossovers, and the problem of relative thermal distributions can also be minimized if one remembers that land/water contrasts are often reversed sometime during the day. All flights for this study were conducted several hours after sunset so as to avoid the problems discussed above.

Equipment Settings

Image quality is influenced by equipment settings used during data collection. At the present time, the dynamic temperature range of most infrared scanners is less than 10°F. Normally, the temperature range is divided into several shades of gray as limited by the recording media. For effective image assessment it is necessary to center the thermal range over the item of interest to insure sufficient thermal detail. Any object outside of the system dynamic range will be recorded as either white or black. Frequently, scenes that include both terrain and waterways will exceed the thermal dynamic range of the IR system. When this occurs, the dynamic range of the sensor must be adjusted to display properly either the land mass or the waterways but not both. Such adjustments are made by setting gain and level controls on the sensor. This is analogous to the contrast and brightness controls of a conventional black and white television receiver. These controls must be set to portray the target temperature within the exposure and contrast range of the recording film.

The IR system used for this study was attended by an operator who monitored the scanner signal on a cathode ray tube (CRT) viewer. The operator insured that the gain (contrasts) and level (brightness) control settings of the sensor were optimum for our purposes. For example, when the objective was the discovery of thermal effluents being discharged into a river, the level controls were set to suppress the low radiance land surface areas and the gain controls were set above normal to provide maximum thermal contrast or sensitivity for river areas. As a result, small temperature differentials in the river were clearly displayed.

Some IR sensors are not operator attended but rather incorporate an automatic gain control feature. The latter technique averages contrast over the entire scene and does not optimize the display for specific objects of interest. Generally, the automatic gain control option is not a desirable feature for applications similar to this study.

The photographic process used to obtain the final positive prints represents another parameter that must be controlled. The direct output of an infrared system is normally negative photographic film. Such negatives were used to produce the positive prints included in this paper. Although some dynamic range was lost in the printing process, printing exposure and processing were carefully controlled to yield optimal tonal rendition of the river thermal profiles. For this report only the overall exposure was varied and no selective dodging or burning-in was used on any of the imagery examples contained herein.

Meteorological Conditions

The effects of meteorological conditions upon the transfer of infrared energy are manifest in both direct and indirect ways. Severe attenuation of infrared radiation frequently can be directly attributed to atmospheric moisture, *e.g.*, precipitation or heavy fog. Incidentally, smoke and moderate haze have very little direct effect on the transmission of infrared energy. The atmosphere also plays a very important indirect role as it influences the heat balance within a scene. Transient wind, humidity and cloud conditions indirectly affect the instantaneous level of infrared energy recorded by sensors for micro-environments within the view of the sensor.

Two rather complex temperature profiles are depicted in Figure 16.[1] These data represent temperature conditions for two 24-hour periods (*i.e.*, on two separate dates) for a single site. The profiles shown are the result of a computer simulation model for a mid-latitude location under two atmospheric conditions: clear sky and heavy haze. The temperatures shown are for the ambient atmosphere above the surface and for depths down to one meter below the surface of the ground. The initial starting time for both days was 4:25 p.m. Two haze cases are shown for the afternoon (1:25 p.m.) of the first day: one with 200 micrograms of particulate matter

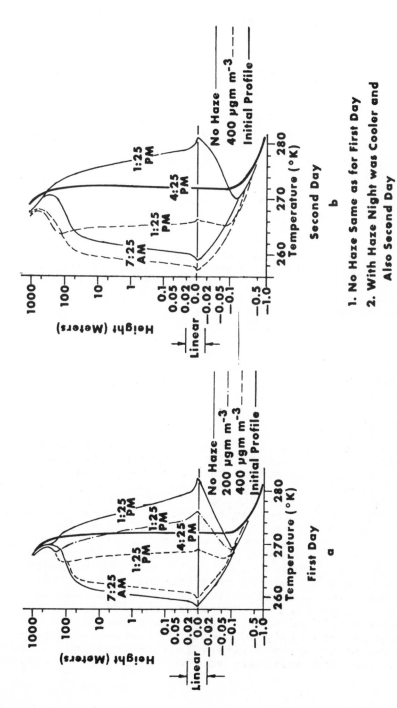

Figure 16. Temperature profiles.

per cubic meter and the other with 400 micrograms.
Surface temperatures for these cases seem to indicate
that a linear increase in haze concentration produces
a near linear reduction in surface temperature. For
this reason the 200 microgram case was not included
for the other time periods. The curves at the left
(7:25 a.m.) show the temperature conditions just
before sunrise.

In summary, solar radiation heats all materials
depending on their absorption and specific heat
characteristics. Winds tend to transfer heat while
fog, dew, and precipitation create a uniform high
emissivity coating on the terrain. The humidity
conditions determine the evaporation rate and the
resulting cooling trend. High clouds, above the
infrared sensing platform, tend to reradiate or
reflect the emitted ground energy and act like a
blanket. An overcast sky slows nocturnal cooling
and some crossovers never occur. The best conditions
for infrared sensing generally are clear, still
nights preceded by fair, sunny days. All data collected for this study were acquired under near
ideal meteorological conditions.

DISCUSSION OF THERMOGRAPH
IMAGERY EXAMPLES

General

Several examples of thermal infrared imagery, or
thermography, collected during this program are presented and discussed in this section. A short
narrative precedes each example and briefly describes
the thermograph. All examples given are positive
prints with the lighter tones representing the warmer
targets and darker tones portraying the cooler ones.
This, however, is a relative condition and there is
no correlation between tonal density and absolute
temperature. Where point radiometric temperatures
have been provided for reference purposes on the
digitally calibrated thermograms, the thermal data
is only approximate. Moreover, precise thermal data
are very difficult to obtain. In summary, the results of this study document the successful application of airborne military infrared sensing systems
to detect and delineate sources of thermal pollution.

Figure 17, Charleston,
West Virginia - (Terrain)

This thermograph depicts terrain detail because the equipment controls were set to emphasize land areas. Note the detail portrayed in the street patterns, highways and foliaged areas. Individual vehicles are discernible along with building detail and bridge structure. The dome of the state capitol building is shown in the upper center portion of the thermograph and a baseball diamond can be seen along the right side of the illustration. Note how difficult it is to see the river barge located in the Kanawha River when the system is set for optimum terrain presentation.

Figure 18, Charleston,
West Virginia - (Chemical Plant)

The level control was suppressed and the contrast amplified for this thermograph so as to emphasize thermal effluents in the river. At least three major and four minor thermal discharges can be seen entering the Kanawha River near the chemical plant. Also note the cooling pond used by industry across the river from the chemical plant.

Figure 19, Charleston,
West Virginia - (Power Plant)

This thermograph clearly shows the effluent discharged by a thermal power plant. Note the current pattern that results when the smaller stream flows into the Kanawha River near the plant. This flow pattern is important if temperature samples are to be made in the river. Also note how the heated outfall hugs the bank and is apparent for some distance downstream.

Figure 20, Washington, D.C.,
Potomac River - (Power Plant)

Thermal pollution is evident in our nation's capital as shown by this thermograph. For example, discharge from a large power plant can be seen as it is swept upstream by the incoming tide. Thermal pollution from this source extends beyond the

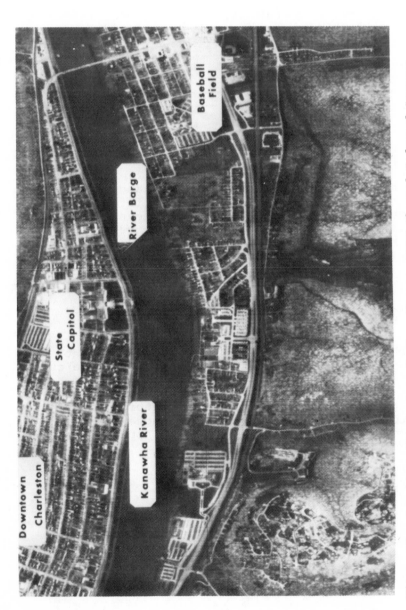

Figure 17.   Location: Charleston, W. Va.   Sensor: USAF Thermal Infrared Scanner. Date: 18 April 1972.   Time: 2100 hrs.   (Air Force Avionics Laboratory)

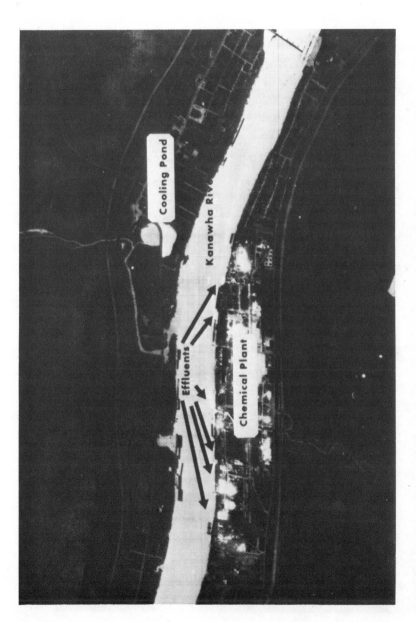

Figure 18.   Location: Charleston, W. Va.   Sensor: USAF Thermal Infrared Scanner
Date: 25 April 1972.   Time: 2015 hrs.   (Air Force Avionics Laboratory)

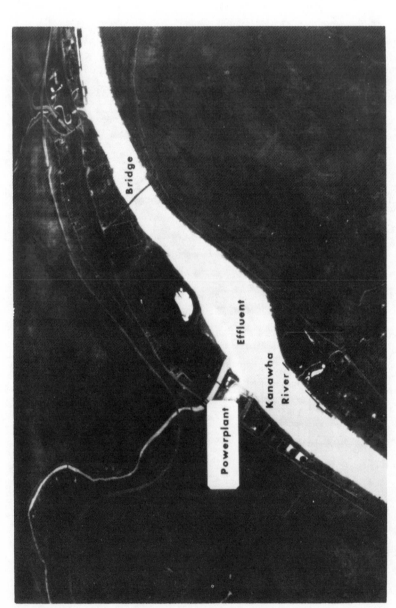

*Figure 19.   Location: Charleston, W. Va.   Sensor: USAF Thermal Infrared Scanner.
Date: 25 April 1972.   Time: 2020 hrs.   (Air Force Avionics Laboratory)*

Figure 20. Location: Washington, D.C. Sensor: USAF Thermal Infrared Scanner. Date: 16 May 1972. Time: 2110 hrs. (Air Force Avionics Laboratory)

Washington National Airport. Exhaust plumes from
several aircraft awaiting takeoff at the airport are
also discernible.

Figure 21, Washington, D.C.
Potomac River - (Power Plant)

This example is essentially the same scene shown
in Figure 20, but it was made on a different heading.
It clearly depicts the thermal discharge from the
power plant, and the extent of the resultant plume.

Figure 22, Pittsburgh, Pennsylvania -
(Three Rivers)

This illustration depicts the confluence of the
Allegheny and Monongahela Rivers as they form the
Ohio River near Pittsburgh. Note that the Allegheny
is somewhat cooler than the Monongahela and that
immediate mixing does not occur. A barge and a tug
can be seen on the Monongahela just above the
junction.

Figure 23, Pittsburgh, Pennsylvania -
(Thermal Effluent on Monongahela River)

Both air and water thermal pollution are depicted
in this thermograph. In this example, both plumes
are emanating from a thermal power plant located in
Elrama, Pennsylvania. Although smoke is normally
invisible to infrared sensors, in this particular
case a high percentage of heated combustion particles are present in the smoke plume, and the
resultant smoke (exhaust) is imaged by the IR sensor.

Figure 24, Pittsburgh, Pennsylvania -
(Thermal Effluent on the Ohio River)

This is another illustration of both air and
water thermal pollution from a power plant. Note
the extent of the smoke plume and the relatively
small water discharge. It appears that a cooling
pond is located near the plant. If this is the case,
the water plume is much smaller than it otherwise
would be. Several barges can be seen in the left
side of the thermograph. White pinpoints of heat
are the stacks of the propelling tug boats.

Figure 21. Location: Washington, D.C. Sensor: USAF Thermal Infrared Scanner. Date: 16 May 1972. Time: 2100 hrs. (Air Force Avionics Laboratory)

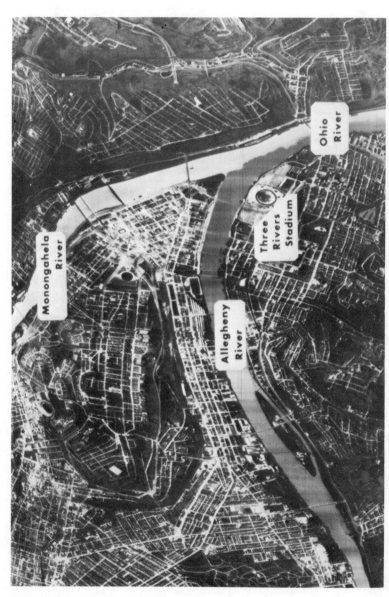

Figure 22. Location: Pittsburgh, Pa. Sensor: USAF Thermal Infrared Scanner. Date: 26 April 1972. Time: 2030 hrs. (Air Force Avionics Laboratory)

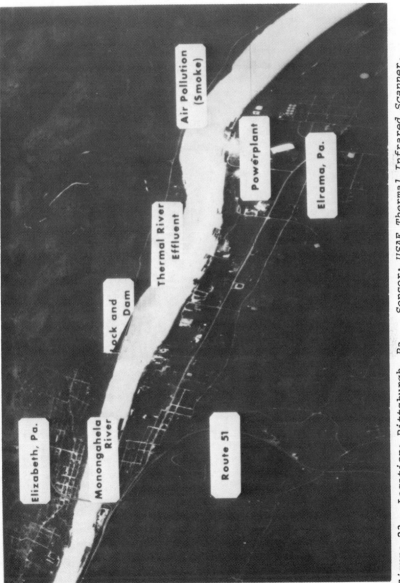

Figure 23. Location: Pittsburgh, Pa. Sensor: USAF Thermal Infrared Scanner. Date: 26 April 1972. Time: 2000 hrs. (Air Force Avionics Laboratory)

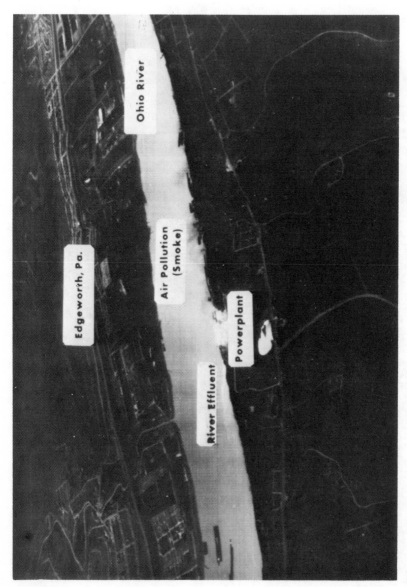

*Figure 24.   Location: Pittsburgh, Pa.   Sensor: USAF Thermal Infrared Scanner.
Date: 25 April 1972.   Time: 2110 hrs.   (Air Force Avionics Laboratory)*

Figure 25, Pittsburgh, Pennsylvania -
(Steel Mill)

This thermograph illustrates the large amount of heat radiated from a steel mill. One major and two minor thermal discharges can be detected where they flow into the Monongahela River.

Figure 26, Pittsburgh, Pennsylvania -
(Thermal Discharges into the Ohio
River at Midland, Pennsylvania)

In this illustration two large and one small thermal plume can be seen in the Ohio River. The Midland steel plant is responsible for two of these while an unidentified industry accounts for the other. Pooling of the effluents near the island is caused by eddy currents in that portion of the river.

Figure 27, Pittsburgh, Pennsylvania -
(Thermal Effluent)

This illustration shows two major thermal discharges emanating from different power plants. The effluent located in the center of the thermograph is quite extensive both in temperature and volume. This imagery was processed to emphasize thermal contrasts in the water.

Figure 28, Pittsburgh, Pennsylvania -
(Thermal Effluent)

This scene is the same as Figure 27, but in this case the image was processed to emphasize terrain details. Note the difficulty in delineating the extent of the river effluents in this thermograph. Also compare the level of terrain detail discernible between this and Figure 27. An original negative contains an extremely large amount of useful information. It is very desirable to have the original negative available and then to process subsequent generations of the imagery so as to enhance the subject of interest.

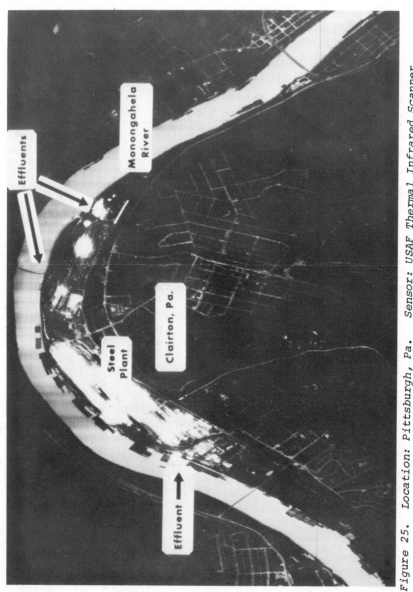

Figure 25.   Location: Pittsburgh, Pa.   Sensor: USAF Thermal Infrared Scanner. Date: 26 April 1972.   Time: 2045 hrs.   (Air Force Avionics Laboratory)

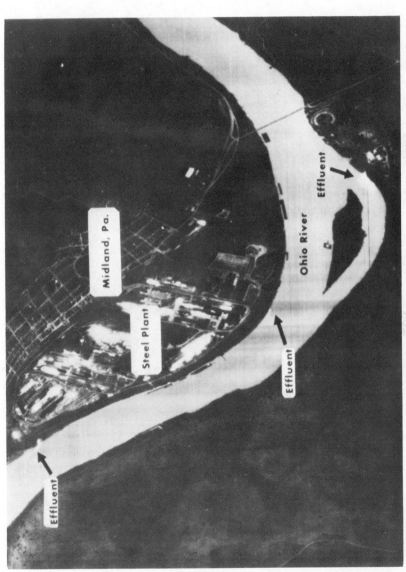

Figure 26. Location: Pittsburgh, Pa. Sensor: USAF Thermal Infrared Scanner. Date: 26 April 1972. Time: 2050 hrs. (Air Force Avionics Laboratory)

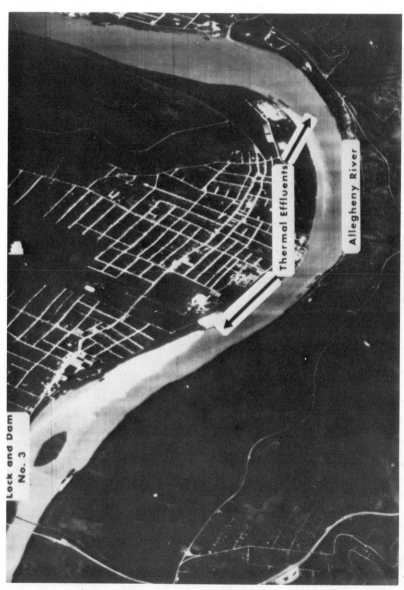

Figure 27.   Location: Pittsburgh, Pa.   Sensor: USAF Thermal Infrared Scanner. Date: 26 April 1972.   Time: 2035 hrs.   (Air Force Avionics Laboratory)

Figure 28. Location: Pittsburgh, Pa. Sensor: USAF Thermal Infrared Scanner. Date: 26 April 1972. Time: 2035 hrs. (Air Force Avionics Laboratory)

Figure 29, Pittsburgh, Pennsylvania -
(Thermal Effluent)

This illustration shows an enlargement of the
largest thermal plume shown in the two previous
figures. This was done to permit direct comparison
with the next three figures, which were collected
using an infrared system with a calibration reference
level. Recall that the thermographs produced by the
USAF thermal infrared scanner are AC coupled;
therefore, the tonal density has little correlation
with the absolute radiometric temperature.

Figure 30, Pittsburgh, Pennsylvania -
(Calibrated Thermal Effluent)

This imagery, showing the same area as Figure
29, was collected by a Reconofax XVI calibrated infrared scanner. It is an analog portrayal of the
radiometric data collected on magnetic tape. Note
that similar levels of detail exist between this
system and the USAF thermal infrared scanner used
for Figure 29.

Figure 31, Pittsburgh, Pennsylvania -
(Calibrated Thermal Effluent)

The calibrated thermal infrared image shown here
depicts spot radiometric temperatures in various
portions of the scene. The central portion of the
thermal plume is 12°F warmer than ambient river
temperature. Note the persistence of the outfall
as it flows downstream. Dispersal does not really
occur until the dam is reached. The river temperature below the dam is 2-3°F higher than ambient
upstream temperature. The annotated digital
thermogram quantifies the degree and distribution
of the thermal discharge.

Figure 32, Pittsburgh, Pennsylvania -
(Calibrated Thermal Effluent)

This illustration depicts thermal contours for
the water surface area shown in Figures 30 and 31.
The area of each radiometric temperature level can
be easily identified and calculated using this
format. Note that spot temperatures are also shown
in this illustration.

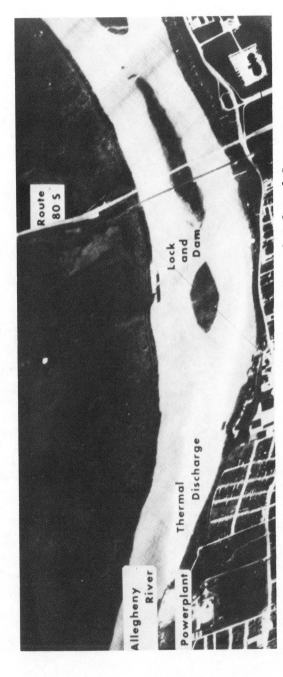

Figure 29.   Location: Pittsburgh, Pa.   Sensor: USAF Thermal Infrared Scanner.
Date: 26 April 1972.   Time: 2035 hrs.   (Air Force Avionics Laboratory)

Remarks: Uncalibrated enlargement of Figures 27 and 28.

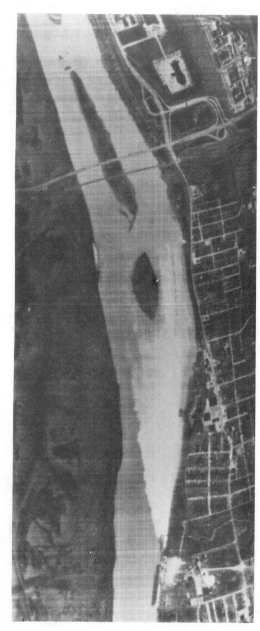

Figure 30. Location: Pittsburgh, Pa. Sensor: Reconofax XVI. Date: 05 May 1972. Time: 2142 hrs. (H R B – Singer Inc.)

Remarks: Analog presentation from a calibrated infrared system.

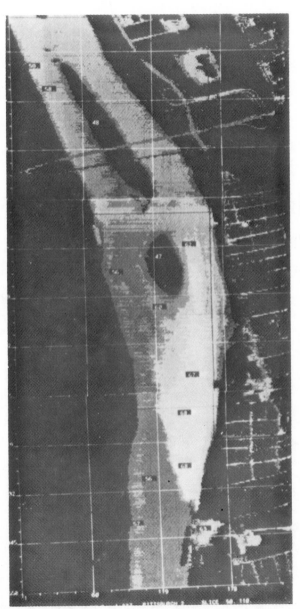

Figure 31. Location: Pittsburgh, Pa. Sensor: Reconofax XVI
Date: 05 May 1972. Time: 2142 hrs. (H R B – Singer Inc.)

Remarks: Annotated digital radiometric record with point temperatures in °F.

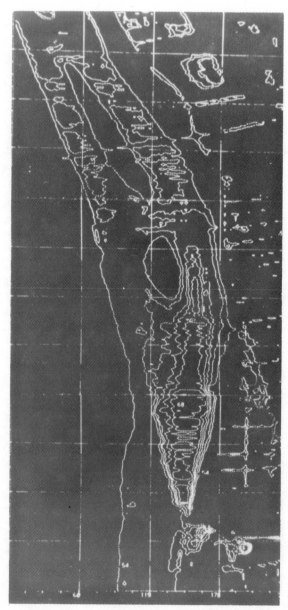

Figure 32.   Location: Pittsburgh, Pa.   Sensor: Reconofax XVI.
Date: 05 May 1972.   Time: 2142 hrs.   (H R B - Singer Inc.)

Remarks: Computer generated thermal contour map in °F.   Contour interval is 4°F from
50°F-62°F and 2°F from 62°F and above.

Figure 33, Gallipolis Ferry, West Virginia -
(High Altitude Terrain Image)

This thermograph, imaged from high altitude,
shows an extremely complex pattern of thermal contours that result in part from elevation and surface
slope. The colder, low lying valleys, shown in the
darker tones, are contrasted with the lighter tones
shown for the higher ground. This phenomenon is
due to micrometeorological conditions that may
exist at a specific time and place.

Figure 34, Wright-Patterson AFB -
(Steam Line)

This thermograph illustrates the use of infrared
imagery for locating and identifying steam lines
found either above or below the surface. Note the
warmer presentation of the buried steam line under
the road and taxiway. This is due to the greater
heat conduction of the overlying concrete as compared with soil. Also note that the elevated steam
line is colder than the terrain background. This
apparent anomaly is due to a heavy layer of
insulation that has a very low emissivity rating.

CONCLUSION

This paper has presented a brief introduction
to infrared theory and technology as it is applied
to remote sensing of the environment. The sample
imagery included in this report demonstrated that
thermal infrared sensors can be successfully used
to determine the location and extent of some
atmospheric and waterway pollutants.

ACKNOWLEDGMENTS

This paper is based upon research conducted by
members of the Reconnaissance and Surveillance
Division, Air Force Avionics Laboratory, Wright-
Patterson AFB, Ohio. Those who made significant
contributions to this study include: Mr. Merle
Carr, Mr. Robert Heuman, Mr. Arthur Arro, and
Mr. Gary Frank.

*Figure 33. Location: Gallipolis Ferry, W. Va.   Sensor: USAF Thermal Infrared Scanner
Date: 18 April 1972.   Time: 2110 hrs.   (Air Force Avionics Laboratory)*

Figure 34. *Location: Wright-Patterson AFB, Ohio. Sensor: USAF Thermal Infrared Scanner. Date: 26 April 1972. Time: 2110 hrs. (Air Force Avionics Laboratory)*

REFERENCE

1. Zdunkowski, W. G. and N. D. McQuage. "Short-term Effects
    of Aerosol on the Layer Near the Ground in a Cloudless
    Atmosphere," *Tellus, XXIV(3)* (1972).

SECTION III

AIR POLLUTION

11.   INTRODUCTION

As seen in the preceding section the control of
water pollution resulting from natural resource extraction is a difficult task.   Controlling air
pollution is equally difficult or more so.   We are
all aware of dirty smokestacks (Figure 35), smog,
and air pollution alerts, especially if we live in
some of the large metropolitan areas of the U.S.

Figure 35.   Power plant emissions, Aliquippa, Pennsylvania.

No one is more aware of air pollution problems than
the industrialist who now has to invest a large
amount of capital in equipment for air pollution
control, equipment that is bulky and very costly
and in many instances rather ineffective, equipment
that rarely means added revenue but almost always
means increased operating expenses, increased process
inefficiency, and added downtime.

Progress in abating air pollution in the United
States has been slow. From an industrialist standpoint, setting an air pollution abatement program
involves the following steps:

1. determining areas with in-plant air pollution
   problems and evaluating the extent of each
   problem
2. determining who has responsibility for air
   pollution control and air quality monitoring
3. setting contingency plans in case of alerts
4. evaluating the process changes and/or added
   equipment needed to bring the plant in compliance with government regulations
5. implementing the air pollution control
   program

Although technology has been utilized in many
industries and power plants, results are sometimes
less than satisfactory. For instance, many authorities favor "tall stacks" for power plants (Figure
36) while many consider this an unsatisfactory
solution and show results to prove it.

Air Pollution in the United States

Air pollution is a severe problem in most metropolitan areas of the United States. Table XVIII
lists the total amount of pollutants released to the
atmosphere from all sources in the United States.
Industrial pollution and fuel combustion from
stationary sources account for one-third of the
total tonnage of pollutants emitted to the atmosphere. These sources represent 96% of the SO_x
pollution, and over half the total emissions of both
NO_x and particulates.

Burning of fossil fuels for industrial purposes
and power generation produce the bulk of SO_x and
NO_x pollution. The iron and steel industries produce 25% of all particulate emissions in the U.S.

Figure 36. *Pollution from the tall stack in the center of the photograph coupled with a temperature inversion clouds up the entire area (Pittsburgh, Pa.).*

Huge amounts of money must be spent by many industries to solve their air pollution problems so that their plants comply with federal, state, and local regulations. An industrialist has four avenues that he can take to abate pollution. These are:

1. use of tall stacks to disseminate air pollutants as much as possible
2. use of control devices
3. process improvement
4. equipment improvement.

The first two alternatives can only be regarded as interim solutions until the last two are implemented. Only process and equipment improvements offer viable hopes over the long range.

Table XVIII

*Estimated Emissions of Important Pollutants
in the United States (1968)*

Source	CO	Particu-lates	SO_x	NO_x	Hydro-carbons
	(numbers represent million tons/year)				
Transportation	63.8	1.2	0.8	8.1	16.6
Fuel combustion in stationary sources	1.9	8.9	24.4	10.0	0.7
Industry	9.7	7.5	7.3	0.2	4.6
Solid waste disposal	7.8	1.1	0.1	0.6	1.6
Miscellaneous	16.9	9.6	0.6	1.7	8.5
Total	100.1	28.3	33.2	20.6	32.0

(Data from "Nationwide Inventory of Air Pollutant Emissions 1968," U.S. Department of Health, Education, and Welfare, Environmental Health Service, N.A.P.C.A. Publication AP-73, Raleigh, N.C. (1968).

The problem that has received the most attention from people in the extractive industries is the control of SO_2 pollution from industry and power plants. This is by far the greatest pollution problem plaguing these industries today. Particulates can be controlled using presently available technology but the technology for the control of SO_2 is still in the development stage. Tieman discusses the progress in SO_2 control technology to date.

The industrialist needs to be concerned with the regulations affecting the operation of his plant. Ganim's paper discusses the federal regulatory machine.

Finally, Gregor's paper illustrates the cyclic nature of a component in the atmosphere such as CO_2 and the effect of man and industry on the amount of this constituent present in various forms.

12. CONTROL OF SULFUR DIOXIDE EMISSIONS

John W. Tieman
Bituminous Coal Research Inc.
Monroeville, Pa.

ABSTRACT

Research on processes to remove sulfur oxides from
flue gas has been in progress for many years. The most
promising removal processes are starting to demonstrate
their special capabilities for controlling different
aspects of the problem. Several processes have been
installed in utility boilers, 100 megawatts or larger,
and include dry additive injection, lime/limestone
scrubbing, magnesium oxide scrubbing, sodium-based
scrubbing, and catalytic oxidation. None of the processes has operated for a sufficient length of time to
be called commercially acceptable. The problems that
have occurred include scaling, erosion, corrosion, and
waste disposal.

INTRODUCTION

The controversy concerning the availability of
commercial processes to control sulfur oxide emissions from power plant boilers is still going on.
State-of-the-art papers on control devices are
presented at practically every technical meeting
held, advising the participants to use low-sulfur
oil, to use low-sulfur coal, to use atomic energy,
to desulfurize the coal, to scrub the flue gases.
Who is right? What should be done? Should we wait
for coal gasification? What about low-Btu gas from
coal, solvent-refined coal, fluidized-bed combustion,

MHD, fuel cells, solar energy, geothermal energy,
atomic fusion? Indeed, there are more questions
than answers.

All of these alternatives to control pollution
from sulfur oxides have been mentioned many times,
of course. Some are short-term but most are long-
term solutions. The purpose of this paper is not
to answer all these questions, but to present a
status report on the various programs being carried
out in an attempt to achieve a solution to this
perplexing and complicated problem.

Research on processes to remove sulfur oxides
from flue gas has been in progress for many years.
Pioneers in this field were the British who, in the
early 1930's, installed a full-scale flue gas
scrubbing unit on a 120-Mw boiler using alkaline
water from a Thames estuary plus chalk. Since that
time, various methods have been studied by many
investigators.

Federal laws covering air pollution started with
the 1955 Air Pollution Control Act. With the advent
of the Air Quality Act of 1967, and subsequent pub-
lishing of criteria and available technology for
control of sulfur oxides in January 1969, the clamor
for information on commercially available control
techniques to remove sulfur from fossil fuels has
rapidly increased.

FLUE GAS DESULFURIZATION

The removal of sulfur dioxide from the flue gas
after combustion has received the most attention as
a short-term solution to the sulfur dioxide problem.
Researchers are investigating a variety of methods,
ranging from the relatively simple injection of
chemical additives to complex chemical processing.
The multiple approaches to desulfurization reflect
both the complexity of the problem and the diversity
of the emission sources.

The volumetric relationship of sulfur dioxide to
total flue gases is of the needle-to-haystack kind,
so desulfurization calls not only for handling an
enormous quantity of gases but also for a chemically
acute process of selection to remove the less than
one-half of one per cent by volume of sulfur dioxide
in the flue gas stream.

Power plants vary widely in size, age, load
pattern, type of location, and characteristics of
fuel used. These are the factors that not only

affect their pollution potential but govern the choice of a practical remedy. The technical and economic feasibility of a desulfurization process can depend ultimately on whether it is used in the right situation.

The most promising sulfur dioxide removal processes are starting to demonstrate their special capabilities for controlling different aspects of the problem. Those processes being tested on a scale large enough to be classified as commercially acceptable include: (1) dry additive injection, (2) lime/limestone scrubbing, (3) magnesium oxide scrubbing, (4) sodium-base scrubbing, and (5) catalytic oxidation. The following definition of acceptability was made by the National Research Council Ad Hoc Panel on Abatement of Sulfur Oxides: "The panel's definition of proven industrial-scale acceptability (*i.e.*, commercialization stature) is satisfactory operation on a 100-megawatt or larger unit for more than a year."

Table XIX presents a summary of the processes in the five categories mentioned, either planned or under test in boilers 100 Mw or larger.

Dry Additive Injection

The flue gas additive program, sponsored by EPA, has now been completed at TVA's Shawnee Station near Paducah, Kentucky. Equipment was installed in 1969 on a 175-megawatt boiler to test the ability of limestone, dolomite, or red mud to control sulfur dioxide emissions. The actual test work was begun in 1970. The entire test program was completed in December 1971 and the equipment has now been dismantled.

The addition of a dry additive to a boiler must be accomplished in such a manner that the reaction between the solid and the gas is favored by optimum temperature, adequate mixing, and proper time for the reaction to occur. Even if optimization of these variables occurs, preliminary tests showed that the maximum sulfur dioxide removal achievable was 50%, with 35 to 40% the norm.

The final results of the TVA tests indicated that limestone distribution is a function of boiler firing conditions. Under given operating conditions, the injection location did not greatly influence distribution. However, a change in operating conditions greatly influenced not only limestone distribution but also velocity and temperature profiles.

Table XIX

Summary of Full-Scale Sulfur Dioxide Removal Systems,
100 Megawatts or Larger, in the United States

Designer	Utility/Station	New (N) or Retrofit (R)	Size, Mw	Date of Operation
Dry Additive Injection				
Tennessee Valley Authority	TVA/Shawnee	R	175	1970-1971
Lime/Limestone Scrubbing				
Combustion Engineering	Union Electric/Meremec No. 2	R	140	1968-1971
Combustion Engineering	Union Electric/Meremec No. 1	R	125	1973
Combustion Engineering	Kansas Power & Light/Lawrence No. 4	R	125	1968
Combustion Engineering	Kansas Power & Light/Lawrence No. 5	N	430	1972
Combustion Engineering	Kansas City Power & Light/Hawthorne No. 3	R	100	1972
Combustion Engineering	Kansas City Power & Light/Hawthorne No. 4	R	100	1972
Combustion Engineering	Northern States/Sherburm No. 1	N	680	1976

Manufacturer	Utility / Plant	N/R	Size	Year
Combustion Engineering	Northern States/ Sherburm No. 2	N	680	1977
Combustion Engineering	Public Service of Indiana/ Gibson	N	650	1975
Babcock & Wilcox	Commonwealth Edison/ Will County No. 1	R	175	1972
Babcock & Wilcox	Kansas City Power & Light/ LaCygne	N	800	1973
Peabody Engineering	Detroit Edison/ St. Clair No. 6	R	170	1973
Research Cottrell	Arizona Public Service	R	115	1973
Tennessee Valley Authority	TVA/Widow's Creek No. 8	R	550	1975
Chemico	Duquesne Light/Phillips	R	100	1973
Chemico	Duquesne Light/Elrama	R	500	1973
Chemico	Ohio Edison/ Bruce Mansfield	N	880	1974
Chemico	Ohio Edison/ Bruce Mansfield	N	880	1974
Chemico	Pacific Power & Light/ Dave Johnson	N	360	1972
Chemico	Arizona Public Service/ Four Corners	N	575	1972

Designer	Utility/Station	New (N) or Retrofit (R)	Size, Mw	Date of Operation
Sodium-Base Scrubbing				
Combustion Equipment Assoc.	Nevada Power/Reid Gardner	R	125	1973
Wellman-Power Gas	Northern Indiana Public Service/Mitchell	R	125	1974
Magnesium Oxide Scrubbing				
Chemico	Boston Edison/Mystic No. 6	R	150	1972
Chemico	Potomac Electric/Dickerson No. 3	R	100	1973
United Engineers	Philadelphia Electric/Eddystone	R	120	
Catalytic Oxidation				
Monsanto	Illinois Power/Wood River	R	100	1972

The amount of sulfur dioxide removal obtained varied considerably and appeared to be a function of the stoichiometric ratio. Approximately 10 to 15% removal was obtained per stoichiometric amount of limestone.

The efficiency of the electrostatic precipitator was also reduced because of the increased ash burden as a result of the limestone addition. Indications are that lime and lime-sulfur dioxide products are not collected as efficiently as coal ash. In addition, these products also affect the ash sluice water quality and must be considered in any decision to install this process.

Despite the obvious advantages of the dry additive process—low cost and applicability to existing plants—the method does not appear usable where a high percentage of sulfur removal is required. However, the process may be suitable in special cases, especially when partial removal is acceptable.

Lime/Limestone Scrubbing

The major emphasis on processes to remove sulfur dioxide from flue gases has been the development of lime/limestone scrubbing systems, as evidenced by the number of full-scale processes shown in Table XIX. The primary reasons for this are: (1) freedom from the problems of marketing a by-product, (2) relatively low cost of raw materials, and (3) high potential sulfur dioxide removal efficiencies. These advantages, however, must be weighed against scaling, erosion, corrosion, and solid waste disposal.

The basic processes used include: (1) injection of limestone into the boiler followed by collection in a wet scrubber, (2) scrubbing the flue gas with a slurry of limestone, and (3) scrubbing the flue gas with lime instead of limestone.

Combustion Engineering

The Combustion Engineering system, an example of basic process #1, involves the injection of limestone or dolomite into the furnace of a boiler where it is calcined to the reactive oxide. The oxides carry through the system with the flue gas to an absorber where, with water, they form a dilute slurry of alkaline hydroxides. Sulfur dioxide reacts both with the oxides suspended in the gas and with the hydroxides formed in the scrubber to produce calcium sulfite and sulfate. At the same time, other particulates are scrubbed from the gas. The solution containing the compounds formed in the reactions drains to the bottom of the scrubber and thence to a clarifier or pump where the solids settle. Clarified water is then available for re-circulation. The clean flue gas passes through a mist eliminator and a reheater before passing to the stack.

Combustion Engineering has installed and operated three full-scale scrubbing systems. Full-scale operating experience of the Combustion Engineering process has been gained at three in-stallations: (1) a 140-megawatt boiler at Union Electric's Meremec Station, (2) a 125-megawatt boiler at the Lawrence Station of Kansas Power and Light, and (3) a 430-megawatt boiler, also at the Lawrence Station of Kansas Power and Light. Two other units, both 100 megawatts in size, at the Hawthorne Station of Kansas City Power and Light were just recently started.

The Meremec system was operated intermittently between September 1968 and June 1971. During this period, a total of 120 in-service days were obtained which included 41 days at half the normal unit load. The problems encountered were many. Severe deposits of reaction products in the boiler convection sec-tion required high pressure jet cleaning. Scaling of the scrubber drain lines, marble bed, and over-flow drains required frequent cleaning. In addition, demister operation allowed particulate and moisture carry-over, resulting in deposits on the induced draft fan rotor.

The decision to discontinue operation of the system was made only after careful consideration of the factors unique to this particular boiler. For example, the boiler design included a tubular air heater and close spacing of the heat transfer sur-faces in the convection section; this caused the excessive deposits that were difficult to remove. In addition, the lack of a convenient by-pass around the sulfur dioxide removal system required reduced load or an outage in case of a malfunction of the removal system.

At the Lawrence Station of Kansas Power and Light, the air pollution control system on the 125-megawatt boiler was started with coal in December 1968. Most of the problems encountered at Meremec were also experienced here. However, a by-pass system had been installed, allowing the boiler to continue operating on a fuel other than coal while repairs were made on the sulfur dioxide removal system. Scaling in the scrubber and other areas required extensive changes to the mode of operation to overcome these problems.

The 430-megawatt unit was activated in the fall of 1971. Operation has been intermittent and modifications were made during 1972 to conform to the operational mode of the smaller unit. Demister washing problems continue to be a source of trouble. During this past winter, the system operated at about 50% availability.

Meaningful data concerning the level of sulfur dioxide removal achieved have not been obtained. Those data that have been taken indicate a range between 50 and 75%, depending on the mode of operation. Dust removal efficiency has ranged from 98.5 to 99.2%.

Both units at the Hawthorne Station of Kansas City Power and Light were activated in late 1972. One of these units, #3, is a boiler injection system; the other uses a "tail-end" process with limestone slurry. Other than mechanical problems, the two systems appear to be performing well.

A 70-megawatt boiler at Louisville Gas and Electric's Paddy's Run Station is scheduled for operation this spring. Carbide sludge, a waste product consisting of calcium hydroxide and calcium oxide from the manufacture of acetylene, will be used as the scrubbing medium in a "tail-end" process.

Babcock and Wilcox

B & W has supplied a limestone slurry scrubbing system for a 175-megawatt cyclone boiler at the Will County Station of Commonwealth Edison. The scrubber system is made up of two identical systems, each taking half of the boiler flue gas. Each system consists of two recirculation tanks, slurry recirculation pumps, a Venturi fly ash scrubber, a sump, a sulfur dioxide absorber, a flue gas reheater, and an induced-draft booster fan. Up to 83% sulfur dioxide removal is anticipated when burning 4% sulfur coal. Operation began in February 1972.

Operating experience for the first year was plagued with mechanical failures, scaling, and demister pluggage. Availability of the two scrubber systems was less than 35% during this period. Sulfur dioxide removal efficiency, based on preliminary test data, ranged from 66 to 96%, depending on the type of operation. The unit is now being operated to solve the major problems of the sulfur removal system without regard to the need for producing power. Once these problems are resolved, there still remains the major problem of sludge disposal, which may, in fact, overshadow all others.

B & W is also supplying a similar type of system for a new 820-megawatt boiler at the LaCygne Station of Kansas City Power and Light. Operation of this unit began in mid-1973.

Peabody Engineering

Detroit Edison is installing a limestone scrubbing system on a 170-megawatt boiler at their St. Clair Station. A Peabody-Lurgi Venturi scrubber will be used to remove fly ash and sulfur dioxide. Completion date: late 1973.

Research Cottrell

Another throwaway system utilizing slurry scrubbing has been developed by Research-Cottrell. They utilize a high energy, flooded disc Venturi for particulate removal and a wetted film, fixed packing absorber tower for sulfur dioxide removal. A demonstration plant is being constructed for a 115-megawatt boiler at Arizona Public Service's Cholla Station. Operation is expected in 1973.

Tennessee Valley Authority

TVA has been studying sulfur dioxide removal systems for a number of years. Recently it was decided to install a full-scale demonstration plant at their Widow's Creek Station on a 550-megawatt boiler. Since no proven recovery methods were available, a throwaway system, using limestone slurry scrubbing, was selected even though such a system involves a solid waste disposal problem. Operation is expected in 1975.

Chemical Construction Corporation

CHEMICO has also been active in studying lime/ limestone slurry scrubbing throwaway systems and has installed many pilot plants. Some of the full-scale units they have installed or are installing remove fly ash now and have the capability of removing sulfur dioxide later. For example:

At Holtwood Station of Pennsylvania Power and Light, a 72-megawatt boiler has been retrofitted with a Venturi scrubber, using water as the scrubbing liquor, to remove 99% of the fly ash. The unit has been operating since late 1970.

At the Four Corners Plant of Arizona Public Service, three boilers, 175, 175, and 225 megawatts each, have been retrofitted with Venturi scrubbers to remove fly ash. With two Venturis connected to each boiler, each unit will be able to operate at 50% load if one of the scrubbers is out of service for repairs or maintenance. The capability exists for sulfur dioxide removal by injecting supplementary sprays in the lower section of each vessel, and by using a limestone slurry instead of water for scrubbing. Operation began in December 1971. Mechanical problems and pluggage are being overcome.

Three Venturi scrubber trains were installed on a 360-megawatt boiler at the Dave Johnson Station of Pacific Power and Light to remove fly ash now. Sulfur dioxide will be removed at a later date. The unit began operation in 1972.

Duquesne Light Company is installing Venturi scrubbers at their Phillips Station and Elrama Station, about 880 megawatts total generating capacity. The Elrama plant will scrub fly ash initially and sulfur dioxide later. At the Phillips Station, one of four scrubbing trains will include a second stage absorber to remove sulfur dioxide. Lime will be used as the scrubbing medium in the second stage system. Both plants plan to start up in 1973.

Ohio Edison recently announced that they will install a sulfur removal system on each of the two 880-megawatt boilers at their Bruce Mansfield Station. The system will be required to operate at 90% sulfur removal efficiency using limestone as the absorbent. Operation is scheduled for early 1974.

Sodium-Based Scrubbing

The problems associated with lime/limestone scrubbing systems have led many investigators to develop sodium-based scrubbing systems to improve scrubber operation. Many of these processes are still in the pilot plant or demonstration plant size. Only two processes have reached the full-scale commercial size, one a throwaway type of system and the other a product recovery system.

Combustion Equipment Associates

Nevada Power is installing a throwaway scrubber system on a 125-megawatt boiler at their Reid Gardner Station. Sodium carbonate will be used as the scrubbing medium. The plant is located in a very dry area and soluble sulfite-sulfate salt can be discarded without causing a major water pollution problem.

Wellman Power Gas

The Wellman-Lord process removes fly ash in a prescrubber, and an absorption column is used to absorb sulfur dioxide with sodium sulfite to form the bisulfite. The bisulfite is sent to a crystallizer where the pyrosulfite is crystallized from the solution. The pyrosulfite is then dissolved in water and sent to the stripping section of the process where sulfur dioxide is removed by steam stripping. Better than 90% of the sulfur is removed.

Following a pilot plant installation at Baltimore Gas and Electric's Crane Station, a full-scale demonstration plant was installed at an 800-tpd sulfuric acid plant, operating on tail gases. Wellman-Lord says that this is equivalent to a 100-Mw boiler installation. The main disadvantage of the process is the need to purge the sulfate formed that cannot be regenerated.

A full-scale demonstration of the process is scheduled by Northern Indiana Public Service at their Mitchell Station. Construction to retrofit a 115-megawatt boiler began in early 1973. Start-up is scheduled for mid-1974. The concentrated stream of sulfur dioxide from the process will be fed to a reduction facility using Allied Chemical's technology to produce high quality elemental sulfur as the final product.

Magnesium Oxide Scrubbing

In some respects, magnesium oxide scrubbing processes are similar to lime scrubbing processes. The principal difference is that the magnesium sulfite and sulfate reaction products can be regenerated, producing magnesium oxide for reuse in the scrubber loop and sulfur dioxide for feed to a sulfuric acid plant or a Claus plant. Three major projects in the United States are underway.

Chemico-Basic

Chemical Construction Corporation of New York and Basic Chemicals of Cleveland have formed a joint company for the purpose of removing and recovering sulfur dioxide from stack gases. Chemico-Basic's main purpose is to promote magnesium-based scrubbing of sulfur dioxide from utility boilers in order to recover the sulfur dioxide in a salable form. The scrubbing is accomplished by an add-on unit using a magnesium oxide slurry for the scrubbing medium. The magnesium sulfite formed is then sent to a separate processing plant for regeneration of the magnesium oxide.

The demonstration plant has been added to a 150-megawatt, oil-fired boiler at Boston Edison's Mystic Station. The spent absorbent is shipped to Essex Chemical Corporation in Rumford, Rhode Island for recovery of the sulfur values in the form of sulfuric acid.

The process began operation in April 1972 and has had intermittent operation during the first year. No major problems have occurred in the scrubbing system itself with regard to scaling and plugging; sulfur dioxide removal has been consistently measured in excess of 90%. The main problems have been in materials handling equipment after scrubbing, and this has been caused mainly by the formation of trihydrate crystals of magnesium sulfite rather than the larger hexahydrate crystals. Thus, the product from the centrifuge was more mud-like than originally anticipated. Dryer operation has continued to be a problem with this type of material. Indications are that a different type of dryer system should be used.

The regeneration step at Essex Chemical has indicated satisfactory operation of the calciner after correction of usual mechanical equipment failures. Approximately 500 tons of 98% sulfuric acid has been produced to date, and magnesium oxide regeneration has exceeded 90%.

A second installation of this process is being made on a 100-megawatt coal-fired boiler at the Dickerson Station of Potomac Electric. Two scrubbers in series will be used, one to remove fly ash and the other to absorb sulfur dioxide. The facility at Essex Chemical will be used for the regeneration of magnesium oxide. Start-up is planned for 1974.

United Engineers

Philadelphia Electric is planning a magnesium oxide scrubbing system on a 120-megawatt boiler at their Eddystone Station. Little information has been made available.

Catalytic Oxidation

The catalytic oxidation of sulfur dioxide to sulfur trioxide and subsequent recovery of sulfuric acid is the basis for the well-known contact process for the manufacture of sulfuric acid. Several investigators have researched this route for application to stack gas clean-up of sulfur dioxide. Only one company, Monsanto, has progressed far enough for a large, full-scale demonstration of the process.

Monsanto Cat-ox

The basis for the Cat-ox process is the high temperature oxidation of sulfur dioxide to sulfur trioxide followed by the recovery of sulfuric acid. Monsanto operated a prototype installation from 1967 to 1969 at the Portland Station of Metropolitan Edison Co. The process has now been retrofitted on a 100-megawatt boiler at the Wood River Station of Illinois Power. Start-up began in September 1972. The prototype tests, 15 megawatt, indicated the capability of the process to remove 85 to 90% of the sulfur dioxide. Efficient removal of fly ash is required to prevent fouling of the catalyst in the converter. The high temperature electrostatic precipitator, ahead of the converter, proved successful in removing over 99% of the fly ash. The information obtained from this operation was sufficient for scale-up to the 100-megawatt boiler size.

For power plant applications, Monsanto offers
two versions of the process: an integrated system
for new plants and a reheat system for use on
existing plants. The two systems differ in the
manner in which the flue gases are treated. In a
new plant, the flue gases leaving the hot precipitator, 850°F, are fed directly to the converter
before going to an economizer and air heater. In
an existing plant, the flue gases leaving the
precipitator are at a temperature of about 350°F
and require reheating to 850°F before entering the
converter. A gas- or oil-fired heater is generally
used for reheating.

Following start-up at the Wood River Plant, many
mechanical problems were encountered. Operation of
the entire unit has only been successful for a period
of hours, with some acid being produced. The major
problem at the present time is the inability of the
reheat system to operate properly. Two induction
heaters are being used but will probably be replaced
by a separate external oil-fired furnace.

The primary advantage of the Cat-ox system is
that no raw materials other than flue gas and catalyst
makeup are required for the process. The size of the
sulfuric acid recovery system, however, will preclude
the use of this process on many plants. The recovery
system is at least as large as the boiler to which
it is attached. In addition, many utilities do not
want to get into the chemical business, but the sale
of the dilute acid (73 to 80%) could help to offset
a portion of the costs. On the other hand, if a
market for the acid were not available, the acid
would have to be neutralized and disposed of, thus
adding to the overall cost of operations.

Other Processes

Many processes are being investigated by companies
other than the foregoing to control sulfur oxide
emissions. Several of these are listed below. Most
are in the small pilot plant or demonstration size
to obtain data for scale-up to commercial size.

TVA

A small pilot plant was installed at the Colbert
Station of TVA to study the limestone slurry scrub-
bing process; ammonia scrubbing has also been studied.
In addition, they have installed a much larger pilot

plant at their Shawnee Station under joint sponsorship with EPA. This latter test facility consists
of three parallel scrubber systems, each capable of
treating approximately 30,000 acfm of flue gas.
The scrubbers selected were the Venturi (manufactured
by CHEMICO), the turbulent contact absorber (manufactured by UOP), and the Hydro-Filter (manufactured
by National Dust Collector Corporation).
    The overall objective of this program is to
evaluate the feasibility and economics of closed-loop
limestone wet-scrubbing processes. The test program
began in 1972 and the major goals are as follows:

1. investigate and solve operating and design
   problems, such as scaling, plugging, corrosion,
   and erosion
2. generate test data to characterize scrubber and
   system performance as a function of the important process variables
3. study various solid waste disposal methods
4. develop mathematical models to allow economic
   scale-up of attractive operating configurations
   to full-size scrubber facilities
5. determine optimum operating conditions for
   maximum sulfur dioxide and particulate removal,
   consistent with operating cost considerations
6. perform long-term reliability testing.

    The entire test program is scheduled for 30 months.
The cost of this installation, including testing,
has been estimated at $8 to $10 million.

Stone and Webster--Ionics

    The first step in the process is to scrub fly
ash-free stack gases with a sodium hydroxide solution
to remove sulfur dioxide. The clean gas is then
passed through a booster fan and reheater and discharged to the stack. The liquid effluent from the
absorber, a mixture of sulfur salts, is fed to a
stripper tower, where dilute sulfuric acid is added
to oxidize all sulfur salts to sodium sulfate and
drive off the excess sulfur dioxide. The sulfur
dioxide is collected and used as feedstock for an
acid plant or solid as liquid sulfur dioxide. The
sulfates pass into a membrane-type electrolytic cell
that generates sodium hydroxide, sulfuric acid,
hydrogen, and oxygen.
    The process has been field-tested in a small
pilot installation at Tampa Electric's Gannon Station.

A demonstration plant is being installed at a
Wisconsin Electric Power plant in a joint project
with EPA. A full scale unit may be built.

Lockheed Aircraft Corporation

Lockheed has devised an electrochemical process
for recovering sulfur dioxide that has gone through
laboratory scale tests only. The process uses no
circulating scrubbing solution and would operate at
temperatures close to those of the flue gases. The
flue gases flow through a chemical bed that removes
most of the sulfur dioxide, which is oxidized in
the reactor to sulfuric acid.

Wade Company

The Wade Company, a division of Ovitron Corporation, has developed a desulfurization process using
a modular-type wet scrubber that simultaneously
removes sulfur dioxide and fly ash from flue gas.
The process emphasizes by-product recovery, and the
choice of reagents to be used in scrubbing is dependent upon the desired end product. A small
pilot plant (3000 cfm of flue gas) has been operated
at New Jersey Power and Light Company's Gilbert
Station. The efficiency of the unit was 98% removal
of sulfur dioxide and fly ash. The scrubbing medium
used was anhydrous ammonia in a water solution. The
company is currently looking for a larger plant
application to obtain engineering design data for
full-scale operation.

Consolidation Coal Company

Consolidation Coal Company has developed a process for removing sulfur dioxide from stack gas.
The process can be applied to existing coal-burning
power plants or engineered into new ones. The
Consol process uses two new previously unexploited
reactions, which involve the use of potassium formate
that is regenerated after recovery of elemental
sulfur.
   In the Consol process, sulfur dioxide is absorbed
by scrubbing the stack gas with concentrated formate
solution at 200°F. Sulfur dioxide reacts rapidly
with the formate to form thiosulfate. The thiosulfate
is then reacted with additional formate at an increased

temperature to yield potassium hydrosulfide. The
hydrosulfide is stripped with carbon dioxide and
steam to produce hydrogen sulfide, from which elemental sulfur is recovered. The formate is
regenerated with carbon monoxide and recycled to
the scrubber.

Following tests in a small pilot plant at Consol's
research laboratories in Library, Pa., a larger unit
was installed at the Cromby Station of Philadelphia
Electric. Operation of the scrubbing system in 1972
was successful and a demonstration-size unit is
planned.

Precipitair Pollution Control, Inc.

This company, a subsidiary of Advance Ross Corporation, has developed a sulfur dioxide removal
process based on use of the mineral nahcolite
(sodium bicarbonate). The mineral is injected into
the flue gas ahead of a filter and converts sulfur
dioxide into a filterable solid. Precipitair has
tested the idea at two different power stations and
has obtained 70% removal of sulfur dioxide.

Atomics International

North American Rockwell's Atomics International
division has developed a flue gas desulfurization
process based on scrubbing the gas with a molten
carbonate solution (a eutectic mixture of lithium,
sodium, and potassium carbonates). Bench-scale
tests of the molten alkali carbonate process show
greater than 95% removal of sulfur dioxide. Unlike
other scrubbing processes, this process does not
cool or saturate the flue gas since the scrubbing
takes place at elevated temperatures. The sulfites
and sulfates formed are reacted with a reducing gas
to regenerate the carbonates and form hydrogen
sulfide. The hydrogen sulfide can be converted to
elemental sulfur or sulfuric acid depending on
marketing conditions.

A pilot plant (10 megawatt capacity) is being
constructed at the Arthur Kill Station of Consolidated
Edison Company. The boiler is oil-fired. The test
program was completed at the end of 1973.

Tyco Laboratories

A laboratory study has been completed on a modified chamber process that would remove sulfur dioxide and the oxides of nitrogen from power plant stack gases and convert them to salable sulfuric acid and nitric acid. The first step of the process would be the oxidation of sulfur dioxide to sulfur trioxide by nitrogen dioxide. This reaction is very fast and requires retention times of only one or two seconds. Gases from the first step will then be passed through a countercurrent packed absorber. Here the gas will be cooled to 150°F, water vapor will be condensed from the stream, and the sulfur trioxide and nitrogen oxides will be absorbed in 80% sulfuric acid as nitrosylsulfuric acid. The dissolved nitrogen oxides will then be stripped from this acid by high temperature flue gases and excess volumes readsorbed in water to yield the nitric acid coproduct.

Princeton Chemical Research, Inc.

The PCR desulfurization process converts sulfur dioxide in flue gas to elemental sulfur. In this process, hydrogen sulfide is reacted with sulfur dioxide at 250-300°F in the presence of a catalyst to produce sulfur and water. Some of the sulfur is recovered and then reacted with methane and water at high temperature in the presence of another catalyst to produce carbon dioxide and hydrogen sulfide for reaction with more flue gas.

Universal Oil Products

In addition to piloting their turbulent contact absorber (TCA) for lime/limestone scrubbing, UOP is installing a proprietary process at the State Line Station of Commonwealth Edison. The pilot plant is reportedly about 22 megawatts capacity, and elemental sulfur will be recovered.

Garrett Research & Development Company

Garrett utilizes a slurry of potassium carbonate in molten potassium thiocyanate as the absorption medium for sulfur dioxide. Following a series of

chemical processes, elemental sulfur is recovered
for sale and the thiocyanate is regenerated.

U.S. Bureau of Mines

The Bureau of Mines has developed a citrate
scrubbing process for the removal of sulfur dioxide
from stack gases. The process is based on the rapid
reaction of sulfur dioxide with hydrogen sulfide in
aqueous solution to form elemental sulfur. The
clean gas is passed through an absorption column
where a bisulfite-citrate complex is formed.
Treatment of this complex with hydrogen sulfide
yields elemental sulfur and regenerates the citrate
ions for reuse. Sulfur dioxide removal efficiency
has been shown to be between 90 and 95%.
A pilot plant, 2,000 scfm of flue gas, is under
construction at the Vigo plant of Pfizer, Inc. in
Terre Haute, Indiana. Peabody Engineering and
Arthur G. McKee & Co. are also involved in the
project.

Monsanto

In addition to the Cat-ox process, described
earlier, Monsanto's Enviro-Chem Systems has developed
the Calsox process for removing sulfur oxides from
flue gas. The process utilizes an organic compound
for absorbing sulfur dioxide. The resulting solution is treated with lime to recover the organic
material and produce calcium sulfate and sulfite
for landfill. A 3,000 scfm pilot plant has been
installed at the Elmer W. Stout Station of
Indianapolis Power and Light to obtain data for
scale-up to a larger size.

Babcock and Wilcox--Esso Research

The B & W-Esso process uses a fixed bed of a
solid sorbent to obtain upwards of 90% removal of
sulfur dioxide from flue gas. The sulfur dioxide
is recovered during the regeneration of the sorbent
and sent to a sulfuric acid plant as feedstock.
A small pilot plant has been operated at a
power plant site. Following preparation of a design
for a 150-megawatt boiler, the system may be installed at the Tanners Creek Station of American
Electric Power.

Westvaco

The Westvaco sulfur dioxide removal process utilizes multistage fluidized beds of activated carbon to absorb sulfur dioxide. The process is carried out at about 300°F. Following sulfur dioxide absorption, the sulfur dioxide is regenerated as elemental sulfur or concentrated sulfur dioxide, depending upon the reductant used. A small, 15-megawatt pilot plant is planned.

Commonwealth Associates, Inc.

Commonwealth has acquired the rights to market the Chemiebau process for sulfur dioxide removal in the Western Hemisphere. Chemiebau acquired the process in 1967 from Reinluft GmbH, and, from observations and operating experiences gained from pilot installations, made significant changes to the original Reinluft apparatus and arrangement.

Basically, the Chemiebau process involves dry adsorption of sulfur oxides by a moving bed or an adsorbent char. The adsorbent is thermally regenerated with a hot scavenging gas and the sulfur dioxide is collected. Since the sulfur dioxide is in a concentrated form, subsequent processing can produce sulfuric acid, liquid sulfur dioxide, or elemental sulfur. The process also claims to remove nitrogen oxides, and in the regeneration step they are reduced to nitrogen and oxygen in the presence of carbon.

The system was piloted in West Germany and England on oil-fired and coal-fired boilers. Commonwealth is in the process of adapting the German design to a modular concept that can be installed at a utility for a demonstration plant.

Envirotech Corporation

Envirotech has developed a double alkali system for the removal of sulfur dioxide from flue gas. This throwaway system uses a highly reactive sodium scrubbing liquor to produce sodium sulfite/bisulfite/sulfate. Lime is then added outside the scrubber loop to precipitate calcium salts, and sodium carbonate is then added for sodium makeup and softening of the scrubbing liquor. The calcium salts are filtered and discarded. The external waste conversion process, separated from the flue gas and

recirculating scrubbing liquor circuits, minimizes
the potential for operating and maintenance problems
by isolating solid wastes from the critical flue gas
flow path.

Envirotech has operated a 3,000 cfm slip-stream
pilot plant at the Gadsby Station of Utah Power and
Light for the past year. Sulfur dioxide removal
efficiency has been about 90% with an inlet sulfur
dioxide concentration of about 400 ppm. Additional
pilot plant testing will be carried out to investigate simultaneous removal of sulfur dioxide and
particulates as well as operation with higher levels
of sulfur dioxide.

A full-scale demonstration plant, 100-200 megawatts, is actively being sought to demonstrate their
system.

General Motors Corporation

GM decided, after considering other possible
alternatives, that a double alkali system was most
practical for their sulfur dioxide control requirements. A pilot plant was built and tested at the
Cleveland Chevrolet Plant. Approximately 3,000 cfm
of flue gas was processed, which represents about
10% of the boiler output when burning 2% sulfur coal.
Sulfur dioxide removal efficiency was about 90%.
Sodium hydroxide was used as the absorbent and lime
was used to remove the sulfite/sulfate from the
solution. Calcium scaling occurred until soda ash
was used for softening.

GM is now in the process of constructing a
full-scale system to handle the entire load at their
Cleveland plant. The load is equivalent to a
generating capacity of 32 megawatts.

FMC Corporation

FMC has operated several pilot plants to obtain
data on their double alkali process. Sulfur dioxide
removal efficiency has ranged from 70 to 95% when
using sodium sulfite/bisulfite/sulfate as the scrubbing medium. Lime is used to precipitate the solid
waste products and sodium carbonate is used as a
softening agent.

FMC is presently negotiating with a large industrial manufacturer in Illinois to install this process
on a system of boilers having an equivalent generating
capacity of 40 megawatts. Operation is anticipated
by mid-1974.

Arthur D. Little/
Combustion Equipment Associates

ADL/CEA are operating a 2,000 cfm pilot plant to test various double alkali configurations. A contract has been signed with the Southern Company for installation of a 20-megawatt prototype system. One half the flue gas from a 40-megawatt boiler at the Scholz Station of Gulf Power Company will be processed. The system will be designed to test double alkali operation with lime only or with lime and limestone. In addition, the system will be designed to test lime and limestone slurry scrubbing. Operation of the facility should begin in mid-1974.

Foster Wheeler/Bergbau-Forschung

Bergbau-Forschung, the central research institute for the West German coal industry, has developed a dry adsorption process for the removal of sulfur dioxide from flue gas. A char developed from non-caking coals is used to adsorb the sulfur dioxide in a slowly downward-moving bed at temperatures between 275 and 300°F. Regeneration of the char is accomplished by washing, which produces a weak sulfuric acid, or by thermal regeneration in a hot sand bed, which produces a concentrated stream of sulfur dioxide. Bergbau has tested the system in a pilot plant having a capacity of about 1,800 cfm of flue gas. Foster Wheeler has been licensed under the Bergbau technology and patents, and has also done some development work on its own.

Foster Wheeler is at present designing an FW-BF system for installation at a major southeastern utility. The adsorber will be capable of handling half the flue gas from a 40-megawatt coal-fired boiler. No date for start-up has been mentioned.

CONCLUSION

From the foregoing discussion, it can be seen that several full-scale sulfur dioxide removal systems are being actively investigated. To date none of the units has shown commercial acceptability, as defined earlier. In spite of this, many utilities are ordering removal systems that have shown the greatest promise. There are, of course, serious reservations on performance, especially with regard to availability and reliability.

A unique characteristic of the power industry is that reliable performance is expected around the clock. Hence the utility is not free to experiment at full-scale and over long periods with untried and unreliable new equipment. Any interruption is a cause of expense, and more importantly, curtailed service will occasion financial loss and some hazards to its customers. Electric power cannot be stock-piled; it must flow smoothly and uninterruptedly to millions of customers.

13. ENFORCEMENT OF AIR QUALITY STANDARDS

Ronald J. Ganim
United States Environmental Protection Agency
Chicago, Illinois

ABSTRACT

Air pollution control involves an interplay of issues
and interests that are dealt with on a daily basis by the
United States Environmental Protection Agency, state pollution agencies, and local authorities. The backbone of
the federal effort is the Clean Air Act of 1970. The
national plan is complemented at the state level by state
implementation plans, which after approval become enforceable by the federal government. Two areas of the Clean
Air Act that are strictly policed at present are "Standards
of Performance for New Stationary Sources" and "The National
Emission Standards for Hazardous Air Pollutants." The
Clean Air Act also gives power to citizens to assume the
role of prevention against polluters violating the Act.
This privilege is rarely exercised.

When I consider the many questions involved in
air pollution control, a thousand issues run across
my mind: the rigid requirements of the *Clean Air
Act*, the unique interests involved in the control of
mobile sources, cost-benefit ratios in any control
enterprise, the complementary federal-state efforts
to enforce air pollution regulations (which are
sometimes not that complementary) and, of course,
the important question of available technology.
These are some of the issues that most regional
offices of the United States Environmental Protection
Agency (EPA) deal with every day. The same questions
are being confronted by state agencies in Ohio,

169

Indiana, Michigan, Illinois, Wisconsin, Minnesota,
and in 44 other states as well as by hundreds of
local agencies: The Cleveland Division of Air Pollution Control, The Chicago Department of Environmental Control and the Lake County Health Department
in Indiana all work toward a common goal: clean
air, breatheable air; air that will encourage life
rather than discourage it.

As an enforcement attorney for one agency out of
many, I share the goal of those working for governments in the air pollution field. But my involvement
and developing expertise is necessarily limited to
certain endeavors. So I leave many of the grand
questions to others who can better explain their
relevance and go on to the area that I know best,
the enforcement of air quality standards.

Before we can understand the present enforcement
policy of the EPA, we need to turn to the backbone
of the federal air effort: the Clean Air Act of
1970. In that year, some relatively innocuous
legislation was transformed into a national program
to significantly reduce air pollution by the middle
of 1975.

The federal scheme begins with the national
ambient air quality standards. Under §109 of the
Clean Air Act, the EPA is required to establish
primary and secondary air quality standards. Primary
standards are those that will protect the health of
the public; secondary standards are those that will
protect the public welfare. Public welfare covers
the effects of air pollutants on soils, water,
crops, vegetation, manmade materials, animals,
wildlife, weather, visibility and climate, damage
to and deterioration of property, and hazards to
transportation, as well as effects on economic values
and on personal comfort and well-being.

The EPA has so far published standards for six
air pollutants. Thus the federal air role is mostly
concerned with achieving air quality standards set
for particulate matter, sulfur dioxide, carbon
monoxide, photochemical oxidants, hydrocarbons and
nitrogen dioxide. The attainment date set by
Congress for the achievement of the primary standards
is mid-1975. Secondary standards are to be achieved
within a "reasonable time." Thus, the major concern
today in EPA regional offices is the attainment of
the primary standards through the control of stationary source emissions by the middle of 1975.

In order to implement the national program,
Congress devised a hybrid federal-state approach.
Each state was to draw up a plan to attain and

maintain the primary and secondary standards at a designated time. These plans are called State Implementation Plans. They were submitted to the Environmental Protection Agency for review, and were approved, or in part disapproved, on May 31, 1972. EPA approval set the clock running for achievement of the primary standards within three years from that date, and for the attainment of the secondary standards within a reasonable time.

The Clean Air Act scheme was a new approach to a national problem. The states were left on their own to devise a plan that would achieve the national standards for the six pollutants, and EPA was bound to approve the plan if it could demonstrate such feasibility. Thus, a state would create its own laws, regulations, and programs to deal with its unique pollution problems, so long as it met the minimum federal requirements. In this way, the national goals and invididual state goals could both be realized.

In their own plans some states decided that both primary and secondary standards should be met at the same early date. Other states stayed close to the federal timetable. The state regulations themselves exhibit a myriad of strategies, tailored by each state to meet both federal requirements and its own interests.

Behind this apparent scheme of individual responsibilities is another interesting provision of the Clean Air Act. Section 113 of the Act gives the U.S. EPA the right to enforce the requirements of a State Implementation Plan. In other words, the federal government can enforce some state air pollution laws. In this way, Congress left more than a supervisory responsibility for cleaning the air of the six designated air pollutants in the hands of the federal EPA.

One other requirement that deserves mentioning is the concept of a "compliance schedule." When Congress mandated a three year clean-up program for the primary air standards, it did not expect the states and EPA to wait until 1975 to see which sources were still violating the law and then prosecute them. Congress required that each State Implementation Plan contain "emission limitations, schedules, and timetables for compliance with such limitations." Thus, a state plan not only needed air pollution laws that would be effective in mid-1975, or earlier, it also needed individual source compliance schedules to assure that individual air pollution sources would be brought under control by

the attainment date through an orderly program.   In
a few words, a compliance schedule is a legally
enforceable clean-up program.

When the State Implementation Plans were submitted to the EPA for approval, compliance schedules
for every air pollution source subject to the primary
standards and not yet meeting them, were required.
However, because of the magnitude of the problem of
developing them on an individual basis with many
sizes and types of industries, these schedules were
not submitted as a part of the State Implementation
Plans.   Today, a large part of the effort in many
state agencies is being directed to developing these
schedules.   As they are completed, and after they
have been subjected to a public hearing, they are
submitted to EPA.   After approval, they become part
of the State Implementation Plan and are enforceable
by the federal government.

With this as a background, let us now take a look
at the direction EPA regional offices are going with
regard to enforcement to guarantee attainment of the
primary standards by mid-1975.

Our goal, mandated by the Clean Air Act, is to
have every significant source of air pollution not
presently in compliance with state air pollution
control laws on a legally enforceable compliance
schedule within a year.   These schedules will
guarantee that the national primary ambient air
quality standards will be met by 1975.   Of course,
as we go forward, we will concentrate our greatest
efforts on the largest sources in the most polluted
areas, particularly those sources that have a
measurable impact on air quality.   However, other
smaller sources will not be ignored; the Clean Air
Act and federal regulations require compliance
schedules for all sources, and the statutory deadline
is a fixed one.

The development of compliance schedules is primarily a state function.   As state-negotiated
schedules are submitted to EPA, federal concern will
shift to other sources.   The outcome will be an incremental program devised by a state or the federal
government for every air pollution source in the
country.

To give some examples of this strategy in action,
the Midwest office of EPA challenged two sources of
air pollution.   The first action involved a secondary
aluminum smelter in Wabash, Indiana.   After gathering
evidence of violations of Indiana's air pollution
control laws, U.S. EPA issued a notice of violation
and confronted the company with the evidence at a

conference provided for under §113 of the Clean Air
Act. At that conference, which was open to the
public, the views of both sides were presented;
there were frank discussions of the technological
problems, as well as a setting up timetable for control of the pollution problem. The result of that
conference was an EPA order, which was in effect a
compliance schedule for the source. And the order
issued had the force of law behind it: potential
penalties for knowing violation of the order range
up to $25,000 per day and one year in prison.

The second case involved the U.S. Steel Corporation's Gary Works and Atlas Cement Plant at Gary,
Indiana. After more than a year of gathering evidence, U.S. EPA issued a notice of violation to the
steel company, and held conferences, which again
were open to the public. The conferences ranged
across every aspect of U.S. Steel's operations that
created an air pollution problem. The requirements
of the Clean Air Act were explained by EPA officials,
and the unique problems of pollution control in the
steel industry were described by the company. The
matter has not yet been finalized, but the outcome
will be an enforceable EPA order.

There are several areas of the Clean Air Act
that will be strictly policed by regional EPA Enforcement Divisions. These are "Standards of Performance for New Stationary Sources" and "the
National Emission Standards for Hazardous Air
Pollutants."

The first category of standards applies to certain new and modified stationary sources and requires
air pollution control that reflects "a degree of
limitation achievable through the application of the
best system of emission reduction which (taking into
account the cost of achieving such reduction) . . .
EPA . . . determines has been adequately demonstrated."
These standards are to be applied to those types of
new sources that "may contribute significantly to
air pollution which causes or contributes to the
endangerment of public health or welfare."   In
December 1971, EPA promulgated national standards
for new or modified fossil-fuel fired steam generators,
incinerators, Portland cement plants, and nitric and
sulfuric acid plants. Any new construction in these
areas must now meet a strict set of federal requirements, and perhaps even stricter state laws. Further,
EPA will soon propose standards for other new sources,
including asphalt concrete plants, petroleum refineries and portions of iron and steel mills.

The second set of standards applies to hazardous air pollutants. These are defined as pollutants that may "cause or contribute to an increase in mortality or in serious, irreversible, or incapaci- tating reversible illness." In April 1973, based on data that pointed to a special health danger for certain substances, EPA named asbestos, beryllium and mercury as hazardous air pollutants, and set standards to protect the public health. The mercury standard applies to mercury ore-processing plants and mercury chlor-alkali plants. The beryllium standard covers many metal-producing operations and some machine shops. The asbestos standard applies to asbestos mills, roadways surfaced with asbestos tailings, certain manufacturing operations and demolition and spraying operations.

The latter two categories of sources were found to be significant contributors of airborne asbestos in urban areas, and thus it is crucial that they be controlled if EPA is to carry out the direction of the Clean Air Act adequately. The demolition stan- dard prescribes procedures that must be followed when almost any building is wrecked. The wrecker or the owner of the building must report his inten- tion to demolish a particular structure to the EPA regional office at least 20 days in advance. The wrecker must follow certain procedures to assure that no asbestos dust is being emitted into the ambient air by wetting down and removing the asbes- tos insulation or fireproofing material before the building is knocked over. The spraying standard is designed to control the use of asbestos spray fire- proofing containing more than 1% asbestos.

For the industrial operations that use and pro- cess these substances, emission standards are prescribed: subject beryllium operations may not emit more than 10 grams of beryllium per day, unless they can show a three year history of compliance with an ambient air standard. Mercury emissions may not exceed 2,300 grams per day. The asbestos standard requires that a source of asbestos either use EPA- approved air cleaning devices or meet a no-visible- emissions standard. Sources unable to meet these standards may apply for a waiver of compliance for up to two years so that they may install the appropriate control equipment.

The enforcement of the hazardous pollutant regu- lations is delegable to the states, and the EPA will soon issue the appropriate guidelines for this.

One final area of concern to the EPA is that of citizen suits. For a long time the courts denied

the ability to sue to citizens who sought to abate
an air pollution source unless they could show a
harm unique to themselves. The question was one of
"standing to sue." Even in the recent case of *Sierra
Club versus Morton*, the Supreme Court held that
Sierra Club did not have an interest in a forested
area in California called Mineral King sufficient to
allow the court to consider the merits of the case.
Today, in §304 of the Clean Air Act, standing
to sue is no longer a question. Any citizen may
assume the role of prosecutor to enforce the Act.
The act states that ". . . any person may commence
a civil action on his own behalf . . . against any
person (including . . . the United States . . .) who
is alleged to be in violation of (A) an emission
standard or limitation under this Act or (B) an order
issued by the [EPA] Administrator or [by] a State
with respect to such a standard or limitation, or
. . . against the Administrator where there is alleged
a failure of the Administrator to perform any act or
duty under this Act which is not discretionary with
the Administrator." There are only two restrictions
on this right to sue: the citizen must give 60 days
notice of the suit to EPA, the state and the alleged
violator, and the citizen is precluded from suing
if EPA or the state "has commenced and is diligently
prosecuting a civil action" in a federal or state
court against the alleged violator. However, in
any such EPA action in federal court, the citizen
may intervene as of right. Further, if the citizen
action involves a hazardous air pollutant or an EPA
order under §113 of the Act, the 60-day waiting
requirement does not apply. For all the obvious
power that §304 gives to the ordinary citizen, it
is rarely used today.

14. MODELS FOR CONTROL OF GLOBAL ATMOSPHERIC COMPOSITION

Bryan Gregor
Wright State University
Dayton, Ohio

ABSTRACT

The only apparent secular change taking place in the
atmosphere as a result of human activity is the accumulation of carbon dioxide from fuel burning. A current model
for this effect is reviewed, and a model is proposed for
the influence of the carbon dioxide increase on atmospheric oxygen. A carbon dioxide spike of two to five times
the present atmospheric level (PAL), tailing back to 1 PAL
over two to five thousand years, should have no measurable
effect on the oxygen reservoir.

INTRODUCTION

Current concern about the atmosphere naturally
focuses on acute pollution by contaminants such as
carbon monoxide, oxides of nitrogen and sulfur,
hydrocarbons, lead, and other particulate matter.
This kind of contamination presents a short-term,
local environmental hazard that lasts until the pollutants fall out or are oxidized, dispersed or otherwise removed from the scene, generally a matter of
days. The contaminants do not accumulate in the
atmosphere, nor on the whole do they seem to have
any measurable secular effect on the other reservoirs
(ocean, biosphere, lithosphere) into which they are
eventually discharged.* The only detectable secular

---

*A well-known exception is DDT, whose accumulation in the
biosphere can be verified by observations in such far-flung
places as Antarctica and Spitzbergen.

change taking place in the atmosphere as a result of human activity is the accumulation of carbon dioxide from the burning of fuel, an effect that contrasts with those mentioned above in being apparently innocuous on the short-term, local scale. It has received considerable attention, mainly in endeavors to predict its effect on the climate and thus, indirectly, on continental ice and the world sea level. Though rather inconclusive (because of uncertainties about the climatic response and about the glaciation process itself), these studies have made good progress in outlining the carbon dioxide balance of the atmosphere and predicting its response to combustion of the fossil fuels.

CARBON DIOXIDE

The atmosphere contains 6×10^{16} moles of carbon dioxide. About 5×10^{15} moles are removed every year by the biosphere (photosynthesis), which pays back almost exactly the same amount by respiration (including oxidative decay). Photosynthesis exceeds respiration by a small differential (4×10^{12} moles per year), which represents fossil carbon that escapes oxidation and is incorporated into the sedimentary rocks, a large reservoir with 10^{21} moles of organic carbon. This sedimentary organic carbon is eventually exposed again by erosion of the rocks containing it. It oxidizes slowly, returning 4×10^{12} moles CO_2 per year to the atmosphere (rate of erosion = rate of sedimentation).

The fossil fuels are a minute fraction (about 0.1%) of this fossil organic carbon, but they contain several times as much carbon as the atmosphere. In the normal course of nature they would be gradually uncovered and oxidized along with the rest over hundreds of millions of years, without affecting the carbon dioxide balance of the atmosphere. Instead we are digging them all up and burning them in a few centuries, injecting into the atmosphere a massive spike of carbon dioxide equal to several times its normal content. The present contribution from this source is 3×10^{14} moles per year; this is expected to double by 2000 A.D. and so exceed 10% of the respiratory product of the whole biosphere. Since coal (the predominant fossil fuel) is expected to last for another 500 years, this will be the approximate duration of the injection phase of the man-made carbon dioxide spike.

The effect is presently visible as an increase
of rather less than 1 part per million per year in
the carbon dioxide content of the atmosphere, which
is now about 315 ppm. This amounts to an annual
increase of 1.5×10^{14} moles, one-half of the man-
made input. The other half must be leaving the
atmosphere by one or more of the natural pathways
that connect it to the remaining carbon reservoirs.
Photosynthesis has already been mentioned. The
photosynthetic activity of plants increases linearly
with carbon dioxide pressure so a carbon dioxide
spike might be absorbed by the biosphere. Even
though respiration would catch up with photosynthesis,
the biomass itself would increase, storing more carbon. The living biomass now accounts for some 8×10^{16}
moles of carbon, and associated dead matter accounts
for about 3×10^{17} more. Assuming that both these
quantities would rise in proportion to photosynthetic
activity, a 10% increase in the latter would result
in storage of more than half the carbon now in the
atmosphere. However the photosynthetic response is
limited by several factors, of which nutrient availability may well be one already prohibiting any
material increase in activity: it is widely held
that marine photosynthesis is now limited in this
way, and the slender nutrient reserve of the tropical
rain forest, which accounts for half of the terrestrial productivity, has recently been emphasized.[1]
The long-term capacity of the biosphere as a buffer
for atmospheric carbon dioxide is therefore open to
serious question.
   The atmosphere exchanges carbon dioxide with the
ocean. The exchange is rapid at the atmosphere-ocean
interface, but between the oceanic mixed layer and
the deep sea it is dependent on advective mixing,
and the mixing time of the ocean is of the order of
1000 years. It is this figure, therefore, rather
than the much smaller one implied by transfer of
carbon dioxide from the atmosphere into the mixed
layer, that will determine the rate at which the
ocean can buffer the man-made CO_2 spike in the
atmosphere. Holland[2] has pointed out that the ocean
itself is buffered by the calcium carbonate on its
floor: a rise in atmospheric carbon dioxide will
finally result in the dissolution of limestone. The
entire atmospheric CO_2 reservoir could be absorbed
in this way by increasing the calcium content of sea
water from 400 to 402 parts per million. The ocean-
carbonate system is thus a very effective buffer for
atmospheric carbon dioxide so long as any massive

atmospheric fluctuations take place on a time scale
compatible with the ocean mixing time.

As seen above, we expect to inject all the
available fossil-fuel carbon, something like 5×10^{17}
moles, into the atmosphere during the next 500 years,
about one-half of the ocean mixing time. Holland[2]
has estimated that this will cause a rise in atmospheric carbon dioxide to 2-5 times the present
atmospheric level and tail back off to 1 PAL over a
period of between 2000 and 5000 years, which, in the
light of the foregoing discussion, seems a reasonable
prediction.

OXYGEN

The atmosphere contains 4×10^{19} moles of oxygen,
so it is in no danger of significant oxygen depletion
by the burning of even 10^{18} moles of fossil-fuel
carbon; this has already been pointed out by
Broecker.[3] The atmosphere receives 5×10^{15} moles
of oxygen each year from the biosphere (photosynthesis), and gives back almost all of this in respiration (including oxidative decay). As with carbon,
there is a balance of 4×10^{12} moles per year in
favor of photosynthesis because of the burial of
fossil organic carbon. The budget is essentially
balanced by oxidation of an equal quantity of fossil
organic carbon exposed by erosion.

The point was made earlier that at least three-
quarters of the biosphere (marine flora and tropical
rain forest) is nutrient-limited or nearly so as far
as photosynthesis is concerned; this would make the
biosphere a much less effective buffer of atmospheric
carbon dioxide than the ocean-carbonate system. The
possibility of some photosynthetic response to the
carbon dioxide spike should not be overlooked, however, when considering the oxygen budget. The
nutrient supply in coastal areas is more abundant
than in the open sea, and the temperate terrestrial
flora may be limited by climate rather than by
nutrient scarcity, and capable of increased productivity under higher carbon dioxide pressures.
Moreover the application of fertilizers to agricultural areas has increased the global transfer rate
of phosphorus from the rocks to the soil (by 60%
according to Lerman[4]), and Mackenzie (personal communication) has determined that man is adding nitrogen
to the terrestrial biosphere in roughly the correct
proportion to complement the phosphorus. It is

therefore worthwhile to estimate whether any increase in photosynthetic activity resulting from the carbon dioxide spike would affect the oxygen reservoir.

The extreme case of a biosphere free from climatic and nutritional limitations can be modeled from observations by Forrester *et al.*[5] on the dependence of photosynthetic rate in soybean leaves on P_{CO_2} and P_{O_2}. Observations of oxygen dependence by Björkman[6] on several other Calvin-cycle plants agree well with Forrester's results, giving some measure of confidence in their application to the biosphere as a whole. The relation between world photosynthetic rate y (in units of 10^{15} mol year^{-1}), x (P_{CO_2} in PAL) and z (P_{O_2} in PAL) is expressed in the following equations[7] (see Figure 37):

$$y = (x + 0.3)\frac{dy}{dx} - 2$$

$$\frac{dy}{dx} = \frac{2}{0.1z + 0.3}$$

$$y = \frac{20x - 2z}{z + 3} \qquad (1)$$

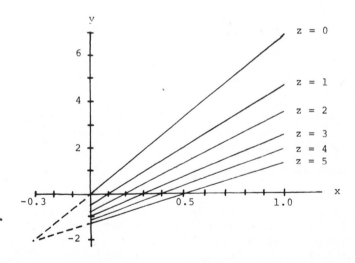

Figure 37. Dependence of world photosynthetic rate y (in units of 10^{15} mol year^{-1}) on CO_2 pressure x and O_2 pressure z, both expressed in terms of their present atmospheric levels.

The oxygen reservoir may be considered as having
a constant output (weathering of fossil carbon) of
4×10^{12} moles per year, and a variable input (from
fossilization of carbon) equal to $10^{-3}y$, *i.e.*,
$10^{12} (\frac{20x - 2z}{z + 3})$. The assumptions made here are that
organic carbon is evenly distributed through the
sediments and the rate of erosion is constant, and
that the rate of fossilization of carbon is proportional to the photosynthetic rate. Given x, this
model can be solved for z as a time series by means
of a computer; however an approximate solution can
be found readily by treating some region of the
curved surface represented by Equation (1) as though
it were a plane. Let k_0 be the slope of $z = 0$, and
let $(y)_z = 0$ be called $y_0 = k_0.x$ (see Figure 38).

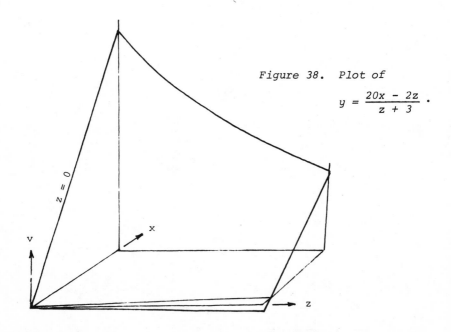

Figure 38.   Plot of

$$y = \frac{20x - 2z}{z + 3} \, .$$

Then $y = y_0 + k.z$, where k is the slope in the direction of the z-axis, here taken as constant. The net
atmospheric gain is now $10^{-3}(y_0 + k.z) = 10^{-3}(k_0.x + k.z)$, and we can write an expression for a box
model of z (the oxygen reservoir):

$$10^{-3}(k_0.x + k.z) \quad \boxed{\quad z \quad} \quad 4 \times 10^{12}$$

Rearranging to treat k.z (which is negative) as a variable output gives:

$$10^{-3}k_0.x - 4 \times 10^{12} \rightarrow z \rightarrow -10^{-3}k.z$$

The solution for z as a function of time is

$$z_t = -\frac{10^{-3}k_0.x - 4 \times 10^{12}}{10^{-3}k} (1 - e^{10^{-3}kt}) \qquad (2)$$

The value of $10^{-3}k$ for the region $0 < (x, z) < 2$ lies between -10^{-7} and -10^{-8} year^{-1}, indicating that about 100 million years would be needed to reach equilibrium in response to perturbations of P_{CO_2}. Assuming Holland's prediction of a carbon dioxide spike rising to between 2 and 5 times the present atmospheric level and falling back to 1 PAL over 2000 to 5000 years to be correct, it is evident that the spike will be over and done with in less than 10^{-4} of the time required for the oxygen reservoir to adjust to it. Its effect on the atmospheric oxygen balance, then, ought to be negligible.

REFERENCES

1. Richard, P. W. "The Tropical Rain Forest," *Scientific American, 229(6), 58* (1973).
2. Holland, H. D. "Control Mechanisms for the CO2 Systems," International Geol. Congress 24th Session, Symposium 117 on Atmospheric Carbon Dioxide (1972).
3. Broecker, W. S. "Man's Oxygen Reserves," *Science, 168,* 1537 (1970).
4. Lerman, A., F. T. Mackenzie and R. M. Garrels. "Modelling of Geochemical Cycles: Problems of Phosphorus as an Example," Work Group on Geochemical Cycles Symposium at Dallas, November 1973.
5. Forrester, M. L., G. Krotkov and C. D. Nelson. "Effect of Oxygen on Photosynthesis, Photorespiration and Respiration of Detached Leaves, I. Soybean," *Plant Physiol., 41, 422* (1966).
6. Björkman, O. In *Photosynthesis and Photorespiration,* (M. D. Hatch, C. B. Osmond, and R. O. Slatyer, eds.) (New York: Wiley-Interscience, 1971), pp. 18-32.

7. Gregor, C. B. and J. Brandeberry.   "Terrestrial Plants
    and Atmospheric Composition (Abstract)," *Trans. Amer.
    Geophys. Un., 54,* 489 (1973).

SECTION IV

RESOURCES OUT OF PLACE

15. INTRODUCTION

Wastes, the discards of civilization, are a by-
product of modern affluence and technology; they are
really resources out of place. One can regard as
waste not only undesirable discards but also land
left derelict after mining has taken place. Reclamation of land and product recycling are attempts to
change barren land into productive acreage and wastes
into new products.

SOLID WASTES

The amount of solid wastes has been growing in
the U.S. in the last few years. Presently, an average
of 10 pounds per day per capita of solid wastes are
discarded. Of these 10 lbs approximately 5.32 lbs
per capita of solid wastes are collected by munici-
palities (Table XX) and 4.68 lbs are disposed of by
individuals and private concerns.[1] A large portion
of these wastes are products made from minerals or
discards from the extraction, production, and processing of minerals and mineral products. In addition, the 10 lbs estimate does not include 7 million
passenger cars, trucks, and buses junked annually in
the United States.
While many minerals and mineral products are
simply discarded, a growing amount of scrap, slag,
tailings, industrial process wastes and refinery
organics are presently being recycled. Secondary
metals are those recovered from discards such as
scrap. Table XXI shows the value of U.S. secondary
metal consumption for some selected metals. As an
added illustration of the importance of the secondary
metals market one can note that gross sales of iron
and steel scrap alone amounted in 1970 to about three
billion dollars.

187

Table XX

Average Solid Waste Collected in the United States

Type of Solid Waste	Urban	Rural	National
	(lbs per person per day)		
Household	1.26	0.72	1.14
Commercial	0.46	0.11	0.38
Combined	2.63	2.60	2.63
Industrial	0.65	0.37	0.59
Demolition, construction	0.23	0.02	0.18
Street and alley	0.11	0.03	0.09
Miscellaneous	0.38	0.08	0.31
Totals	5.72	3.93	5.32

(Data from U.S. Department of Health, Education and Welfare for 1968).

Table XXI

Consumption of Secondary Metals in the United States

Metal	Quantity Consumed in 1967		Dollar Value
Copper	1,243,000	tons	$1,096,000,000
Aluminum	885,000	tons	403,000,000
Lead	554,000	tons	161,000,000
Zinc	263,000	tons	76,000,000
Antimony	25,568	tons	26,000,000
Tin	22,790	tons	71,000,000
Mercury	22,150	flasks	12,000,000
Silver	59,000,000	ounces	101,000,000
Gold	2,000,000	ounces	70,000,000

Note: Dollar values were calculated by the Bureau of Mines
Office of Mineral Resource Evaluation using prices posted
in American Metal Market, May 7, 1969. The price of gold
has increased three fold since that time. The price of the
other metals has also varied slightly.

Even though a growing portion of solid wastes
from resource industries is being recycled, the
largest proportion still is simply dumped in landfills and waste piles around the U.S. These are
truly resources lost, out of place.

LIQUID WASTES

Liquid wastes account for a growing portion of
the pollution problems in our rivers and lakes. Resource industries contribute liquid wastes in all
phases of their operations, from extraction to product
finishing.

When petroleum is extracted, brine is sometimes
produced in large quantities. Wood, in the first
paper of this section, notes that in 1963 about 360
billion gallons of brine or nearly one billion gallons per day was produced with oil in the United
States alone. When the crude oil is transformed
into gasoline and other by-products in a refinery,
numerous substances, organic and inorganic, go into
the liquid waste stream and are discarded. As noted
by Reno in this section, some but not all products
can be recovered from the liquid waste stream.

Another liquid waste resulting from resource
extraction is "acid mine drainage." As noted earlier
in the book, in abandoned coal mines where pyrite is
exposed to air, rain water chemically reacts with the
pyrite to form sulfuric acid and $FeSO_4$. This highly
acidic runoff is a dangerous liquid waste that can
increase the acidity of bodies of water and literally
destroy the ecosystem.

GASEOUS WASTES

Gaseous wastes include all the gases that come
out of industrial smokestacks as well as gases and
suspended particulates emanating from nonpoint sources.
Gaseous wastes greatly contribute to our air pollution
crisis.

Burning of fuels accounts for 31.4% of nationwide
particulate emissions and industrial processes account
for an additional 26.5%. Ore smelting, refinery
operations, coke-making, and manufacture of sulfuric
acid account for 21.4% of SO_2 emissions in the United
States. Industrial processes as a whole account for
10% of nationwide emissions of carbon monoxide and

15% of hydrocarbon emissions. To prevent these
emissions, particularly from point sources, industry
has been spending growing sums of money to install
control equipment to transform these gaseous wastes
into solid or liquid form. Little has been done to
reclaim gaseous wastes.

LAND RECLAMATION

In many instances, resource operations have been
guilty of leaving behind barren areas where nothing
will grow. These areas are one of the worst waste
products that we have. More recently a lot of pressure has been exerted to force these companies to
rehabilitate such lands.

Surface mining in the U.S. accounts for 52% of
the nation's coal production and a sizeable percentage
of the production of many important minerals. Unreclaimed land resulting from surface mining operations
includes excavation sites, spoil areas, water reservoirs, mineral processing plants, and waste disposal
areas.

Many factors need to be analyzed to restore a
parcel of land properly to serve useful needs. Elzam
describes in a paper in this section some of the work
being done to reclaim some of Ohio's strip-mined lands
in the limestone-poor parts of the state where acid
formation is most intense.

The Secretary, U.S. Department of the Interior,
has assigned high priority to the reclamation of
surface-mined lands. At his request a national conference on the reclamation of surface-mined lands for
recreation was held in Washington, D.C. on May, 1973.
Regional conferences were also held on this subject
subsequent to the national meeting.

Rehabilitating surface-mined lands for recreation
is an excellent way of bringing them into productive
use. The coal industry as well as the sand and gravel
industry are making great progress in this respect.
Higgins[2] has reported that a study by Southern
Illinois University recently has shown that of 166,000
acres surface-mined for coal in Illinois as of 1971,
approximately 40,000 acres are presently utilized for
outdoor recreation. Of these 11,800 acres represent
water bodies, 97% of which have pH above 4.0.

REFERENCES

1. Black, R. J., A. J. Muhich, A. J. Klee, H. L. Hickman, and R. D. Vaughan. "The National Solid Waste Survey; An Interim Report," U.S. Department of Health, Education and Welfare, Cincinnati, Ohio (1968).
2. Higgins, T. W. "Reclaiming Surface-Mined Coal Lands for Outdoor Recreation," Proceedings of the Regional Conference on Reclamation of Surface-Mined Land for Outdoor Recreation, Kent State University, Kent, Ohio (1973), pp. 54-64.

16. USE OF UNDERGROUND SPACE FOR WASTE STORAGE THROUGH INJECTION WELLS

Leonard A. Wood
United States Geological Survey
Reston, Virginia

ABSTRACT

Although oilfield brine has been injected into saline aquifers for many years and more than 40,000 brine-injection wells exist, injection of industrial waste through wells did not begin until after World War II and only about 250 waste-injection wells had been drilled by 1972. Emphasis on cleaning up streams and lakes accelerated the construction of waste-injection wells in the 1960's. Implementation of the Federal Water Pollution Control Act and the possibility of increased state regulation of deep waste injection may cause either a rapid increase or a leveling off in construction of injection wells.

Properly sited, constructed, and operated, waste-injection wells can provide safe storage of industrial wastes. Aquifer plugging by particles in waste, or by chemical reactions among waste, rock, and formation fluid, or by growth of bacteria, can prevent the operation of an injection well. Environmental problems may result from injection of waste. The waste displaces brine that may flow to the surface or to fresh aquifers through abandoned test holes or old wells that have been destroyed by corrosion. Also, excessive injection proessure may fracture confining beds and permit waste to flow into other aquifers. Geological exploration, hydrologic testing and evaluation, and engineering, design and construction will reduce both the possibility of well failure and the chances of pollution.

Publication of this paper authorized by the Director, U.S. Geological Survey.

The use of waste-injection wells is a much smaller threat to the environment and the pollution of ground water than is the improper surface disposal of waste, or the careless handling and storage of chemicals, accidental spills of petroleum products and other liquids, and many other activities unrelated to waste storage.

INTRODUCTION

The residual waste from many present day industrial processes and even from some waste-treatment processes is toxic and not subject to ordinary methods of treatment. Such wastes are hazardous if released to streams or allowed to percolate into the shallow subsurface environment. Faced with the high cost of further treatment or recycling, many waste-producing industries are either trying to or are investigating emplacement of waste effluents in deep underground "space." The use of underground space for waste storage is an alternative that quite properly should be considered, provided enough is known about what the waste will do and where it will reside until it is no longer a threat to the environment.

First, we should define what we mean by underground space. I will restrict my remarks to natural underground space, which is the interconnected pore or fracture space in rocks into which a fluid can be emplaced. As McKelvey[1] pointed out, "underground space is not *vacant* space, because, like other kinds of earthly space, it is already occupied by a liquid or gas that must be compressed or displaced if the space is to be used for storage." Most natural underground space is made up of the voids between sand grains; less abundant is the space in fractures between blocks of rocks. Some underground space is the result of solution of limestone, and although individual openings can be large, space created by solution is much less in total volume than intergranular and fracture space.

When storage in underground space is mentioned, most persons think of deep injection wells. However, most fluids get into underground space by percolating from the land surface in a manner similar to infiltrating rain water. Most ground water contamination present in the United States was caused by liquid percolation from waste disposed on the surface or in shallow pits or from other shallow waste disposal activities such as septic tanks. In many ways, wastes

percolating from the land surface are more of a
threat to fresh water supplies than are injection
wells, largely because most fresh water is within a
few hundred feet of the land surface. By sheer
numbers, the waste disposal points on the surface
are a greater threat to the environment than
injection wells.

Injection wells first came into general use in
the oilfields where disposal of brine produced with
oil and gas became a problem after local, state, or
national legislation prohibited disposal of the brine
in drainage ways or in pits. Under that earlier
practice, the brine seeped into the subsurface and
contaminated shallow ground water. Although much
of the injected brine is returned to the same horizon
from which it came in order to maintain pressure
(waterflood) in the oil field, a larger part of the
brine is injected in other horizons for disposal
purposes only. Thus, the brine not used for water-
flooding is injected into aquifers other than those
yielding the oil and brine, often into a much shallower
aquifer and perhaps the first one that contains saline
water.

Accurate, up-to-date figures on the amount of
brine produced in oilfields and disposed of in this
way are not available. However, the Interstate Oil
Compact Commission[2] published data on oilfield-brine
production for 1963. These data were summarized in
an article in *Petroleum Engineer*[3] in July 1967, which
is the source of the data in Table XXII.

Table XXII

Production and Disposition of Oilfield Brine,
Associated with Oil Production in the United States, 1963
(in million gallons per day)
Compiled by Interstate Oil Compact Commission

	Quantity	Per Cent of Total
Produced	994.6	100.0
Disposition		
Injected for waterflood	328.5	33.0
Injected for disposal	385.7	38.8
Unlined pits	117.5	11.8
Impervious pits	0.9	0.1
Streams and rivers	43.3	4.4
Miscellaneous	118.8	11.9

In 1963, about 360 billion gallons (8.6 billion
barrels) of brine, or nearly one billion gallons
(23.7 million barrels) per day, was produced with
oil in the United States. About 72% of the brine
was injected through wells into either the oil
horizon from which it came or into some other
horizon.

The number of wells in use for oilfield brine
injection in 1973 is large but unknown. Warner[4]
quoted an estimate by the Texas Water Development
Board of 20,000 such wells in Texas alone in 1965.
Others have estimated that there are at least as
many brine injection wells in other oil-producing
states combined, so we have an estimate of 40,000
or more for all of the United States in the mid-1960's.

Many factors have served to change the amount of
brine produced and the method of disposal in the last
10 years. "No-pit" orders and more aggressive enforcement by the states have eliminated many of the
so-called evaporation pits that were in use in 1963,
and may have increased the amounts of brine injected.
This also may have eliminated some of the brine
production by shutting down some of the marginal oil
wells. Then, too, increased demand for oil has
eliminated restrictions on oil production, and brine
production undoubtedly has increased along with oil
production. Furthermore, the amount of brine produced per barrel of oil may have increased as greater
production stress is placed on the oilfields. On the
other hand, the more restrictive policies recently
adopted by some of the states may influence the amount
of brine produced and injected. Illinois, for example,
now protects all ground water that contains less than
10,000 mg/l (milligrams per liter) of dissolved
solids from industrial waste injection.[5] Other
states have limited the bottom-hole pressure in the
injection well to prevent hydraulic fracturing of
the disposal zone or the confining layers.

Public Law 92-500, named the Federal Water Pollution Control Act which is the charter under which
the Environmental Agency administers water pollution
control, excludes oilfield brine from the term *pollutant* (Sec. 502 6 B) if it is "injected into a well
to facilitate production of oil or gas," or if it
is "derived in association with oil or gas production
and disposed of in a well," and "if the well used to
facilitate production or for disposal purposes is
approved by authority of the state in which the well
is located, and if such state determines that such
injection or disposal will not result in the degradation of ground or surface-water resources." Oilfield

brine injection will be influenced by how the state
regulations are written and enforced (with the approval of the Administrator of the Environmental
Protection Agency). One predictable result is that
general rules and regulations governing brine injection should become more uniform nationwide.

Widespread concern for the quality of water in
streams and lakes led industries to a search for
alternate methods of waste disposal and many have
seized upon the injection methods developed by the
oil industry. Although a few industrial waste injection wells were in use during the forties and
perhaps even before, Donaldson[6] inventoried only 30
wells as recently as 1964. Warner[7] illustrated the
growth in numbers of industrial waste injection
wells from one or two in 1950 to 110 in January 1967.
The Interstate Oil Compact Commission listed 118 in
1968,[8] and Warner[9] inventoried 246 waste injection
wells in mid-1972. About 200 of those inventoried
by Warner[10] had been constructed in the period 1964-
1972. About two-thirds of the 246 wells were being
operated in 1972. About one-tenth of the wells were
either plugged and abandoned after operating for
some period or were considered failures at the start.
Of the remainder, half are rarely operated or are on
standby and half were being made ready for operation
at the time of the inventory.

Four out of five industrial waste wells in use
in 1972 were operated by chemical, petrochemical,
refinery, or natural gas plants or by pharmaceutical
companies. Many of the wastes are brines although
most contain other substances such as acids, metal
salts, and organic compounds.

Some important differences between the injection
of oilfield brine and the injection of industrial
waste make the waste injection well more complicated
to operate. The typical oilfield-brine injection
well conducts a natural brine into an aquifer, sometimes the same aquifer it came from, or into one
that is usually not greatly different. If returned
to the same aquifer, then the brine partially re-
stores the pressure depleted by oil and brine production. The typical brine injection well averages
a few gallons per minute. For example, in Texas,
injection wells averaged about 330 bbl per day per
well in 1963, or about 277 mgd was injected into
20,000 wells and the average rate was about 10 gpm.
On the other hand, the industrial waste well injects
a complex fluid that will probably react with either
or both the formation fluid and the aquifer rock
skeleton. The waste from some plants is different

at different times because the plant may produce a batch of one product for a while, then change over to a batch of another product. About half of the industrial wells accept from 50 to 400 gpm and about one-sixth accept more than 400 gpm. However, the number of wells receiving less than 50 gpm is increasing. Special precautions in preparing the waste are necessary to prevent plugging of the injection zone. The waste well nearly always adds to the existing natural formation pressure.

The oilfield-brine injection well is usually in or on the edge of an oilfield and extensive data on the depth, geochemistry, porosity, and permeability of the injection zone are readily available to the operator. But, the industrial waste injection well may be miles from the nearest well that penetrated to the target injection zone and few data on the characteristics of the zone are available until a test well is drilled.

Certain problems may be encountered or caused by either an oilfield-brine injection well or an industrial-waste injection well. For discussion purposes, I will separate the possible problems into two categories, functional and environmental.

Functional problems relate to success or failure of the well to deliver the brine or the waste to the injection horizon. Failure of casing, packers, the cement envelope, or other hardware in the ground will shut down the well, if repairs are uneconomical. This kind of failure can be prevented by proper design and construction. A plugged aquifer will prevent fluid entry into it at satisfactory rates and safe pressures, and may also cause a well to be abandoned. A common cause of plugging is particulate matter not filtered from the injection fluid beforehand. Recharge experiments with water from city systems of drinking water quality have shown that such water often contains enough particulates to significantly reduce aquifer permeability in the well bore in a few hours to a few days.

Chemical reactions in the well can also cause aquifer plugging. The reactions can occur between wastes in the injection stream, between wastes and the formation fluid, and between wastes and the aquifer materials. Any or all of these reactions can produce precipitates that will plug the aquifer in the well bore or in the vicinity of the well. Geological and geochemical studies should be made prior to injection so that chemical reactions can be predicted and prevented or controlled. Addition of chemicals at the surface treatment facility or

preceding the injection fluid with a buffering fluid
may prevent chemical plugging.

Of course in some cases the chemical reactions
tend to increase aquifer permeability in the vicinity
of the well bore. For example, an acid waste injected into a carbonate aquifer may dissolve the rock
and enlarge solution channels appreciably.

Aquifer plugging by biological growths is not
well understood. Bacteria are in the air, in the
water, in the rocks, and perhaps in the waste. Some
will thrive in an environment that is hostile to all
others. Certain bacteria may thrive near the well
bore if oxygen is dissolved in the injection fluid.
Other bacteria may thrive in an aquifer after the
injected fluid becomes anaerobic by chemical reactions as the fluid moves away from the well bore.
Corrective measures can only be successful after the
infestation has been diagnosed.

Another functional problem results when an injection well is constructed in an aquifer that is not
permeable enough to accept the fluid as fast as it
is introduced. Pressure rises rapidly to the maximum
allowable and the injection rate must be reduced to
prevent exceeding safe limits. This problem causes
the well to be classified as a failure unless a lower
rate can be tolerated or unless the well can be re-
worked and completed in another aquifer. This
problem can be prevented, or at least predicted, by
proper hydrologic testing before injection.

Environmental problems are those outside the well
and, although they may be caused by either failure
of the well or by conditions in the vicinity of the
well, they may not physically prevent normal operation, *i.e.*, fluid may be injected at normal rates
and pressures and damage may not be evident, at least
for a long time.

Fresh ground water may be contaminated by injected
fluid flowing upward from the injection zone or from
a leaky casing resulting from faulty design or construction, material failure, or inadequate cement
around casing. The injected fluid might also find
its way into another resource if aquifer plugging
causes pressure to build up unchecked until the seal
around the casing is ruptured, or in extreme cases,
the confining layer is fractured.

A more common impact that may persist undiscovered
results from the presence of abandoned test holes and
wells, either uncased, cased with corroded casing, or
improperly plugged. Normal pressure build-up in the
injection aquifer displaces formation brine upward
through the holes. The displaced brine may move into

a fresh aquifer for long distances before it reaches
a well or other discharge point where it can be observed. Or, the brine may move upward to the land
surface causing the abandoned well or test hole to
flow salt water. Commonly, any casing in such
abandoned wells is corroded at the surface and
discharge continues just below the land surface.
The phenomenon is commonly incorrectly identified
as a "salt spring."

Another possible environmental impact caused by
increased pressure in the injection zone is the upward movement of displaced brine through a supposed
confining bed that was not adequately described either
by geological studies or more directly by test drilling. Lenticular shale beds cannot be depended on to
restrict movement of either injection fluid or
displaced brine.

Legal use of underground space for waste storage
may pose more difficult problems than either the
construction of wells or the environmental impacts
of waste injection. Underground space is treated as
property subject to lease in the case of natural gas
storage.[1] If similar space is used for storage of a
noxious waste, the space is lost to the owner.
Should he be compensated for it? Unlike underground
natural gas storage sites where a structure is sought
to retain the gas so that it might be recovered,
waste more commonly is injected in flat lying or
gently dipping beds that underlie much of the country.
If structural control is lacking, liquid waste would
move with the brine in the aquifer, even after injection has ceased although perhaps more slowly.
Sometime, sooner or later, the waste is going to
cross property lines and trespass under another
person's land. What are his rights. Frank Trelease[11]
points out, in his interesting paper in American
Association of Petroleum Geologists Memoir #18, that
the waste disposer is using space that the owner of
the surface might effectively use for disposal of
waste or other purposes. The courts no longer hold
that ground water moves in mysterious ways unknown
to man.

The Environmental Protection Agency (EPA) has
generally followed the policy announced on October 15,
1970 (by their predecessor, Federal Water Quality
Agency, FWQA). That policy opposed "the disposal
or storage of wastes by subsurface injection without
strict controls and a clear demonstration that such
wastes will not interfere with present or potential
use of subsurface water supplies, contaminate interconnected surface waters, or otherwise damage the

environment." The policy also stated that "where
subsurface injection of wastes is practiced, it will
be recognized as a temporary means of ultimate disposal to be discontinued when alternatives enabling
greater environmental protection become available."
E. T. Conrad[12] reported in January 1973, that the
EPA is drafting a new statement that is very similar
to the 1970 statement, but which will contain more
detailed guidelines for injection wells than were
in the 1970 policy statement. The number of new
industrial-waste injection wells drilled in the
future may be curtailed if EPA policy is implemented
by regulations. Although the individual states now
regulate injection wells, PL 92-500 (Sections 301,
303, 402) requires each state to prepare and implement a waste management plan for all kinds of waste.
The plan must be approved by the Administrator of
EPA within a specified time limit. PL 92-500 is not
clear on whether waste-injection well regulations
must be part of the approved plan. Neither is it
clear how the EPA policy will be interpreted. Recent
public statements by EPA seem to endorse disposal of
secondary treated sewage into the "boulder zone"
wells in certain coastal cities in Florida rather
than construct long expensive ocean outfalls or
construct plants that will give the sewage tertiary
treatment.

Successful and safe waste injection can be
achieved at many locations, if geologic and hydrologic
conditions are properly evaluated, and if engineering,
construction, and operation are given proper attention.
Each injection well and waste represent a unique
combination and each must be evaluated individually
and conjunctively to anticipate and prevent problems.
T. D. Cook[13] points out in his foreword to American
Association of Petroleum Geologist Memoir #18,
"Successful applications are numerous. They did not
in the past, nor will they in the future, occur
accidentally. They require knowledgeable personnel,
a background of research and planning of the entire
project, a dedication to execute each step as skill-
fully as possible, a determination to observe and
monitor results, and a commitment to alter or halt
a program which presents possible dangers to man
and his environment."

REFERENCES

1. McKelvey, V. E. "Underground Space--An Unappraised Resource," in *Underground Waste Management and Environmental Implications,* Am. Assoc. Petroleum Geologists Memoir #18 (1972), pp. 1-5.
2. Interstate Oil Compact Commission. "Water Problems Associated with Oil Production in the United States," Interstate Oil Compact Commission, Oklahoma City, Okla. (1965).
3. "Crack Down on Oil Field Pollution," *Petrol. Eng., 39(7),* 33 (1967).
4. Warner, D. L. "Deep Well Injection of Liquid Waste," Public Health Service Publication No. 999-WP-21, Robert Taft San. Eng. Ctr., Cincinnati, Ohio (1965).
5. Ives, R. E. and G. E. Eddy. "Subsurface Disposal of Industrial Wastes, First Supplement," Interstate Oil Compact Commission (1970).
6. Donaldson, E. C. "Subsurface Disposal of Industrial Wastes in the United States," U.S. Bureau of Mines Inf. Circ. 8212 (1964).
7. Warner, D. L. "Deep Wells for Industrial Waste Injection in the United States," Federal Water Pollution Control Admn. Publication #WP-20-10, Cincinnati, Ohio (1967).
8. Ives, R. E. and G. E. Eddy. "Subsurface Disposal of Industrial Wastes," Interstate Oil Compact Commission, Oklahoma City, Okla. (1968).
9. Warner, D. L. "Survey of Industrial Waste Injection Wells," NTIS, vol. 1 - AD-756-641, vol. 2 - AD-756-642, vol. 3 - AD-756-643, Springfield, Va (1972).
10. Warner, D. L. Personal communication.
11. Trelease, F. J. "Liability for Harm from Underground Waste Disposal," in *Underground Waste Management and Environmental Implications,* Am. Assoc. Petroleum Geologists Memoir 18 (1972), pp. 369-375.
12. Conrad, E. T. and N. E. Hopson. "Future Demand for Underground Wastewater Disposal in the United States," paper presented at the Water Resources Conference, American Society of Civil Engineers, Washington, D.C., January 29-February 2, 1973.
13. Cook, T. D., Ed. *Underground Waste Management and Environmental Applications,* Am. Assoc. Petroleum Geologists, Memoir #18 (1972).

17. HANDLING REFINERY WASTES

Gordon Reno
Shell Oil Corporation
Deer Park, Texas

ABSTRACT

Actions taken to abate all forms of pollution at the
Houston Refinery of Shell Oil Company are discussed.
Currently, only garbage and trash are disposed of off-
site. An enviable position has been achieved chiefly by
taking advantage of economic opportunities and refinery
alterations to reduce the effect of the refinery on the
environment.

INTRODUCTION

An aggressive day-to-day approach minimizing the
effect of refinery operations on the environment is
in action at Shell's Houston Refinery. Key to its
success is the firm resolve of the refinery's
management that the job will be done.
The problem can be analyzed by using an analogy
to the fire triangle. In the case of the fire
triangle, removal of one leg eliminates the problem
(Figure 39). Conversely, in my analogy removal of
one leg insures that an environmental problem will
result. I have depicted the refinery management
team in the place occupied by heat or source of ig-
nition in the fire triangle. Absence of a strong
resolve by local management insures that environmental
problems will result. By its nature the operating
department is the fuel for an environmental problem
and is so located. At Houston Refinery, responsibility for avoiding environmental problems and for

203

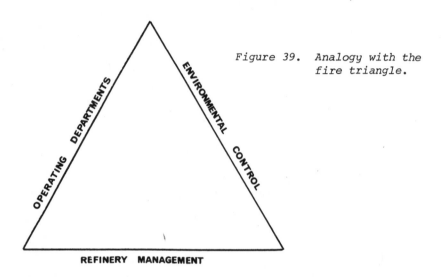

Figure 39.   Analogy with the fire triangle.

quick response if they occur is placed firmly on the operating department. Environmental Control occupies that leg of the triangle assigned to oxygen in the fire triangle. This function is a policing one whose key target should be avoidance not penalty.

DAY-TO-DAY MONITORING AND RESPONSIBILITY

Air

Operating departments are expected to operate all facilities effectively, making operating changes and responding to upsets in a manner to avoid violations of environmental standards. In the event an upset does result in a potential environmental problem, the department is expected to notify the Environmental Conservation Department and take immediate steps to minimize the impact and eliminate the source of the problem. Shutdown plans must include steps to avoid violations of environmental standards.

Individuals noting environmental problems, such as darker than normal combustion emissions or in-plant odors, make a report (by telephone) to the gateman on duty who relays it to the Environmental Conservation Department for action. The source and elimination of the problem is promptly sought whether its origin is in the refinery or at a neighboring plant. Correction of the problem or progress is communicated to the individual initiating the report. A full report is prepared for circulation to management. Complaints from outside the refinery are handled by the same format. This procedure, in use for over 20 years, has led to many solutions by focusing attention on repetitive problems.

Texas air regulations are directed mainly to source control. Hence the Environmental Conservation Department conducts only minimal ambient monitoring. The department is responsible for records confirming source control.

Water

The Environmental Conservation Department tests for compliance with permits and runs tests to anticipate problems from plant operation that will affect wastewater treatment. When a problem is noted, the departments that might be involved are alerted.

Operating departments are expected to respond promptly to a known problem affecting the wastewater treatment facility. They are also expected to report problems as they originate and take action to moderate and correct the problems. In a few cases monitoring is carried out to avoid repetitive problems. Examples are sulfide content of desalter water and pH of wastewater from a sulfuric acid alkylation plant. Shutdown plans must include procedures to avoid upsetting the water treatment plant.

Oil Spills

The Dispatching Department has the responsibility for detecting spills regardless of source (unknown, tankers, or shore facility). Having noted a spill, they proceed to report the spill, stop the discharge, contain the spill, and arrange for clean-up. The dock foreman is delegated authority to hire the clean-up contractor without delay.

AIR POLLUTION ABATEMENT

Houston Refinery has achieved a high degree of air pollution abatement by taking timely advantage of opportunities to conserve natural resources. In some cases a profit has resulted.

Since 1947 several steps have been taken to eliminate open burning with objectionable smoke generation. In 1947 mercaptans were recovered for sales via solutizer treating. Recovery of crude oil emulsions was begun in 1949. In 1953 we placed in service a method of including acid sludge from sulfuric acid alkylation with spent acids returned to a vendor for refortification. Full recovery of oil/water emulsions was initiated in 1959. Open burning was essentially eliminated by 1960. Burnable trash has been removed by a contractor since 1961. In 1961 the use of soil cultivation, a method of destroying the oils in oily solids from refinery clean-out operations, was begun. Results of a research project, partially funded by EPA, demonstrating this process are now available from the U.S. Government Printing Office.

A unit designed to recover 50 T/D (tons per day) of sulfur at approximately 90% efficiency was streamed in 1952. A second unit of 30 T/D capacity was added in 1966. In 1970, two 150 T/D units designed for 96% efficiency were added as part of a significant refinery capacity increase. Included in the expansion project was almost total removal of H_2S from refinery fuel gas using diethanolamine treatment and a two-stage sour water stripper, which removes ammonia from the H_2S fed to the sulfur recovery plants.

Liquid knockout for the main refinery flare was provided in 1954, recovering valuable hydrocarbons. The flare was made smokeless in 1958 and equipped for gas recovery in 1962. Two more flares were equipped for smokeless burning in 1967. A new smokeless flare chiefly for hydrogen-producing and -using plants was installed in 1970. Our last flare was equipped for smokeless burning in 1972. Continuous efforts directed to minimal flaring have proved profitable.

In 1963, an energy recovery package was installed on the Catalytic Cracking Unit of the Houston Refinery. An expander turbine recovering about 9,000 HP from regenerator flue gas was included. To protect the turbine, an effective third stage catalyst separator was needed, following two stages of conventional cyclones. A feed preheater was installed to recover sensible heat from regenerator flue gas plus heat of combustion of carbon monoxide. In 1967, a fourth stage separator was added to recover fines from the third stage separator to meet a 40% opacity regulation effective in 1968. An in-stack opacity meter was made operational in 1968. The facility

without modification will meet 1974 Texas Air Control
Board regulations promulgated in early 1972.

Steps have been taken to eliminate odors and/or
hydrocarbon emissions. Floating roofs were provided
for sour crude oil storage and gasoline storage
facilities. Since 1952, as discussed earlier, we
have had an organized response to odors detected,
resulting in correction of many problems. A key to
elimination of fugitive hydrocarbon losses, usually
odorous, was the switch to mechanical pump seals,
which began 20 years ago.

Meeting requirements of the Clean Air Act of
1970, the Texas Air Control Board passed regulations
for several pollutants on January 26, 1972, requiring
compliance by December 31, 1973. As a result of the
actions discussed above, Houston Refinery found itself in compliance except for floating roofs for
storage tanks containing products with 1.5-11 psia
actual vapor pressure.

A substantial floating roof program for tanks
storing these products is now nearing completion.
A favorable contract for the effort was achieved by
advanced planning leading to timely initiation of
the project.

WATER EFFLUENT CONTROLS

Facilities insuring that water leaving Houston
Refinery will not harm the environment are a practical
blend of the old and the new. By timely and regular
changes, the refinery has been a leader among
Houston Ship Channel industries in providing improved
water discharge. Even though performance is excellent, incremental changes continue.

Process water receives the full treatment, *i.e.*,
oil removal, pH adjustment, and two-stage biological
treatment. Our trickling filter placed in service
in 1963 was the first in a refinery on the Houston
Ship Channel. The biological facility was converted
to two-stage by addition of an activated sludge unit
in 1970. The latter addition was one of several
changes that resulted in a reduced pollutant discharge
concurrent with a significant increase in refinery
throughput.

A key to the refinery's successful pollution
abatement is to be found in the operating areas.
Here, a combination of attitude and equipment designed to avoid sending potential pollutants to the
treating plant is evident. We estimate only 10-20%

of potential pollutants reach the treatment plant.
Included in these defensive facilities are a two-
stage sour water stripper, chemical recovery sumps,
and minimum discharge of cooling water.

Another key to success is the separation of
waters containing little or no pollutants for lesser
degrees of treatment prior to discharge. Inclusion
of these waters in the treating system would decrease
its effectiveness.

WATER POLLUTION ABATEMENT

In 1946 some moves at storm water segregation
were made, and the forerunner of our demulsifying
activity became operational.

In 1947 the recovery of mercaptan for sales
abated water pollution in addition to its elimination
of open burning. Installation of water draw boxes
for tankage increased separation of storm and process
water. A pond for deoiling treating clays was placed
in service in 1949. Included in a lube plant that
became operational in 1949 was a sump for recovery of
phenols, avoiding pollution at the source. This was
the first of several similar facilities that are key
features of our abatement program.

In 1950 Houston Refinery placed in operation the
first Sour Water Stripper in a Gulf Coast Refinery,
removing hydrogen sulfide and ammonia from the wastewater discharge. The gases were directed to the
Sulfur Recovery Unit in 1952. Control of pH by
operating departments was started by the Alkylation
Department in 1950 and by the Utilities Department
in 1953 when they began control of pH of water from
demineralizer regeneration.

Settling ponds were provided in 1954 to improve
oil recovery from the main process sewer. Distribution baffles were added in 1956. In 1963 the settling
ponds became retention ponds prior to biological
treatment. A biological oxidation pond was provided
in 1954 to treat sewage. This facility is still part
of our treatment system today.

Sales of treating caustics for recovery of
phenols and sulfides was initiated in 1956. This
removes a significant biological load from the water
treatment facility. Processing of ship ballast water
via a pond system separate from the main refinery
waste stream made oil removal more effective in 1958.
Demulsifying of oil/water emulsions begun in 1959
resulted in further improvement.

Soil cultivation, mentioned earlier, began in
1961 and diverted some pollutants from the sewer
system. A most significant step was the installation
in 1963 of a trickling filter. Results from this
facility, which included a final clarifier to remove
solids, surpass the performance of some more modern
facilities. Chemical flocculation was added in 1967
to improve water quality further.

Location of a new oil-water separator, which was
needed in 1967, required that the deoiled water be
pumped to the biological treatment facility. Taking
advantage of this, a portion of the water was diverted to cool hot-oil drop out from thermal cracking
units, thereby saving purchased water.

In 1969 as part of the refinery expansion project,
a 50,000 bbl ballast water tank was added to further
improve treating of ballast. A two-stage sour water
stripper replaced the single-stage unit in service
since 1950 and more effectively removed sulfides and
ammonia from the sewer system. Reuse of ammonia-
bearing water by Shell Chemical reduced the volume
of water to the treating system of the refinery.

In 1970 as part of refinery expansion, addition
of an activated sludge basin downstream of the
trickling filter converted Houston Refinery's waste
treatment to two full stages of biological treatment.
The same year, the lagooning pond, which serves as
ballast water treatment, was increased in surface
from 2 acres to 20 acres. Also included in the expansion project were chemical sumps and substantial
air cooling. Examples of materials for which sumps
were provided are diethanolamine and sulfolane.
Use of air coolers obviously reduces need for cooling
water.

In 1971, water discharge from the soil cultivation area was diverted to the refinery process sewer
for biological treatment. A request to operating
departments for better water management to improve
our ability to handle waste loads resulted in a 500
gpm drop in flow through the refinery process sewer.
By careful monitoring, this decrease has been
maintained.

During 1972, two projects were completed that
improved storm water handling. One of these, on a
storm sewer, provided for better oil separation under
severe flow conditions. The other combined protection of the biological facility from high tides, a
problem caused by land subsidence in the area, with
increased water retention and one-stage treatment
of extreme storm flows prior to discharge.

Flow reductions in 1972 came from increasing
cooling water concentration and a concerted effort
to avoid loss of steam condensates. Water discharge
averaged 30 gallons per barrel of crude in 1972.

OIL SPILLS

Prior to late 1971 oil spill cleanup at our dock
was handled by our own forces. Late that year we
and eight other Houston Ship Channel industries
formed a cooperative to supplement our ability to
respond to any large spills. Since most members had
an in-house capability, their choice was to supplement this capability with a contractor who was required to have available a specified minimum amount
of equipment and be prepared for a rapid response to
an oil spill. As a member, when an oil spill occurs
we now have the choice of using our own in-house
capabilities or calling on the contractor. If the
problem should be beyond this capability we can
borrow additional equipment from other members. An
added feature of the cooperative is that the contractor is available to nonmembers including the
Captain of the Port, U.S. Coast Guard, who must
respond to spills of an unknown source.

SUMMARY

An enviable position has been reached chiefly by
taking advantage of economic opportunities and refinery alterations to reduce the effect of the
refinery on the environment. Our aggressive and
continuing program to find opportunities to improve
our performance has been a factor in the results
achieved. Insistence by refinery management that
all operations minimize their environmental effects
has been valuable.

18. STREAM WATER CHEMISTRY OF MODEL RECLAMATION OF STRIP-MINED LAND

O. E. Elzam
Department of Biology
Case-Western Reserve University
Cleveland, Ohio

ABSTRACT

In recent years strip mining for coal moved to the sandstone-rich and limestone-poor regions of southeast Ohio. As a consequence, acid conditions far exceeding those known in the past prevented revegetation of the land by the usual practices. In the past year, a field experiment on approximately 60 acres was constructed to deal with new ways of reclaiming such lands. The experiment involved:

1. burial of all shale and coal-related strata as deeply as possible within the overburden
2. applying a sandstone cover on top
3. returning top soil to the land
4. terracing to prevent erosion
5. planting vegetative cover.

During the experiment, water outflow, ground water, pond water, and well water were regularly collected and their chemical composition monitored, In addition, plant growth was observed and charted. An evaluation of environmental consequences of the reclamation scheme is presented.

INTRODUCTION

Strip mining for coal in the United States and other parts of the world has intensified in the past decade, and from all indications it will be increased aggressively in the near future.[1-4] The reason is the lack of other proven and easily available energy resources. The lack of resources has created an energy crisis requiring extraordinary measures. For example, on September 9, 1973, the President of the United States urged the relaxation of air pollution standards so coal could be used instead of the cleaner energy sources (*e.g.*, oil). Coal also can be converted to gas, which is another energy resource in short supply. These developments indicate that strip mining the vast coal reserves of the U.S., with their relatively high quality coal, will be substantially increased in the near future rather than curtailed.

With the assured increase of strip mining activity, it is in the national interest to discover new revolutionary and successful land reclamation procedures. In the past, reclamation of strip-mined land varied from state to state and region to region. Consequently, there was and is a great variability in the quality of reclamation.[1,5] These great differences occur mainly because there are only a few states with standards for reclamation of strip-mined land. The lack of standards may be because there are no proven and successful guidelines and procedures for reclamation that have been thoroughly researched and documented.

The general impression of the public is that strip mining in all situations and cases is detrimental to the ecology, biology, hydrology, and aesthetics of the affected land.[1,5-7] In Ohio, because of the complete lack of experimental field work, strip mining for coal and subsequent land reclamation were little understood. Recently, in an extensive study by Case-Western Reserve University sponsored by N.S.F., it was found that the effect of strip mining on the environment was vastly different in different regions of southeast Ohio.[5] The study indicated that when the overburden of the strip-mined land contained substantial amounts of limestone, there was little acidity or heavy metals found in streams leaving the land and, consequently, revegetation of the land was found to be fast and complete. On the other hand, when the overburden of the strip-mined land was predominantly sandstone, the streams originating in the affected land were acid and revegetation of the land very difficult. In the latter case, further work indicated that correction of the problem would be costly and economically unattainable.[1]

The lack of scientific work dealing with reclamation of strip-mined land in general and the need to find suitable reclamation procedures in the sandstone-rich coal regions of southeast Ohio led to this cooperative work with Consolidation Coal Co., Central Division of Cadiz, Ohio. The work presented in this paper deals with reclamation of land in the sandstone-rich overburden in Guernsey County, Ohio. Reclamation work was initiated prior to the strip mining operation, and actual land reclamation took place soon after coal removal. It was found that the model reclamation steps followed in this work led to land rejuvenation successful revegetation, and elimination of acid discharge to streams.

MATERIALS AND METHODS

General Site Description

The experimental strip mine site of 60 acres is located in Oxford Township, Guernsey County, Ohio, approximately one mile north of Fairview, off Township Road 61. The site and surrounding natural hardwood forest is owned by Consolidation Coal Company, Central Division of Cadiz, Ohio.

The topography is hilly with noticeable mudslides and downhill land movement.[1] Strip mining of the Pittsburg #8 seam began in June 1971 and was completed in May 1972. The coal analyzed was 12,400 Btu and 5.11% sulfur on a dry basis.[8] The geology of the site is described in U.S.G.S. Professional Paper 380 entitled "The Geology and Coal Resources of Belmont County, Ohio." One feature to be mentioned here is the complete lack of limestone in the overburden and the predominance of sandstone. The few inches (2-3") of top soil on these hills was held in place by the hardwood forest vegetation. The pH of the soil was acid (4-5) and productivity described to be low. Farming and cattle raising in the past in this region were marginal and even lumber production was unprofitable.[1]

Winters in the area are very cold and summers are hot and humid. There is little snowfall in the winter and the rainfall is rather evenly distributed throughout the year, with summer being the driest.[9] Precipitation contribution from heavy thunderstorms amounts to a large portion of the 35" yearly average.[10]

Land Reclamation

Land reclamation started on June 1, 1972. At first, the potentially acid-forming exposed coal and coal-related strata and some of the sandstone rocks were buried as deeply as possible. Then, the less acid-forming sandstone rocky strata was distributed evenly throughout the experimental sites forming three hills of about 20 acres each. The resultant topography was that of gentle slopes, thus satisfying the new State of Ohio reclamation guidelines. As a result there were no high walls or very steep slopes to be encountered in the three experimental sites.
The three hills were then treated as follows:

Treatment 1. Top soil (a mixture of all horizons a, b, and c), removed from the land prior to strip mining, was evenly distributed on this site to thicknesses of 1.5 to 2 feet. The resultant soil layer was somewhat compact and the gentle slopes were smooth and even throughout.

Treatment 2. Terraces were built along the resultant land contour at 50 to 100 feet intervals. The terraces were about 10 feet wide and 5 feet deep. No top soil was introduced to this site. Terracing was built on this hill for the following reasons:

1. to stop land erosion by reducing water velocity on the land.
2. to stop acid formation through saturation of the land with water and the exclusion of air (O_2) needed in pyrite oxidation:[11]

$$FeS_2 \xrightarrow[H_2O]{O_2} FeSO_4 + H_2SO_4$$

3. to increase the amount of water available for plant use and growth.

Treatment 3. Terracing was built as in experimental hill 2. However, a top soil dressing of 1.5 to 2 feet was added to this site.

The three combinations of treatment mentioned
above were designed to establish the most suitable
reclamation procedure that would result in the
elimination of stream acidity and stream salt loading
toxic to life, in the reduction of land erosion,
and in the fastest land revegetation. Upon completion of land preparation, it was necessary to hand
seed the experimental sites because of the heavy
and constant rain during the month of June 1972.
Otherwise the land would have been disked and seeded
with a heavy drill. Plants introduced to all
experimental sites were:

1. yellow sweet clover 25%
   [*Melilotus officinalis* (L)]
2. birdsfoot trifoil 25%
   [*Lotus corniculatus* (L)]
3. orchard grass 50%
   [*Dactylis glomerata* (L)]

Germination was good even though a very large percentage of the seeds were washed away by the heavy
rains.

Land, Water, and Plant
Sampling Procedures

Stream water originating from the three experimental hills was sampled prior to reclamation and
weekly thereafter for about one year. Stream water
originating in nearby natural hardwood forests was
sampled frequently for comparison purposes. In
addition, stream water originating from other
partially reclaimed strip-mined land in the same
vicinity was sampled for demonstration purposes.
Ground water, well water, and pond water
bordering the experimental hills were routinely
sampled to assess the effect of strip mining on the
water quality of the land adjacent to the experimental
sites.
Rain water was collected in equipment built
similar to those described by Likens *et al.*[12] to
determine its quality. Total rainfall, in addition
to intensity and frequency, was monitored by a continuous rain gauge monitor (Bendix Corporation).
Plant material and soil samples were collected
several times during the year for chemical analyses.
All samples, either water, plant, or soil in
origin, were immediately refrigerated at 5°C and
brought to the laboratory. In the laboratory they
were kept at 4°C.[13]

Chemical Analysis

Water Chemistry

Within a few hours after water collection (3 to 4 hours at most), the samples were first filtered through 0.45 millipore filters and then the pH, acidity, and specific conductance measured.[13] Aluminum, iron, manganese, and sulfate were analyzed within 24 to 48 hours after water collections. Other metals such as calcium, magnesium, sodium, and potassium were analyzed no more than 3 to 4 days after collection. Appropriate measures were followed to avoid ionic interferences in analysis when using the Perkin-Elmer atomic absorption spectrophotometer (model 303). A complete documentation of the analytical procedures used have been described in detail elsewhere.[5,13]

Other elements such as zinc, copper, mercury, and cadmium were determined in our samples. However, since their concentrations were so low and insignificant it was decided not to follow them on a regular basis.

Soil Chemistry

Saturation extracts of soils collected from the three experimental hills were obtained (designated as soil solution available for uptake by plants) and chemical analysis performed as described above for water analysis.[13,14]

Plant Chemistry

Plant samples were collected from the three experimental hills, refrigerated, and brought to the laboratory where they were first thoroughly washed with distilled demineralized water and then fresh weights taken. Then, the plant material was dried in an oven at 60°C for one week. After dry weights were taken, the samples were pulverized in a Willey Mill to obtain homogeneous samples.[15]

The dry plant material was wet ashed and the samples brought to a known final volume.[16] Chemical analysis proceeded as described for water analysis.

RESULTS AND DISCUSSION

Stream Water Chemistry

The main objective of this investigation was to find a solution to the acid streams runoff from strip-mined land that is so frequently reported for the sandstone-rich coal regions of southeast Ohio.[1,5,17,18] The monthly pH averages of all streams originating in the three experimental hills is reported in Figure 40 for a one-year period. The symbols that appear in Figures 40-45 are shown in the caption to Figure 40.

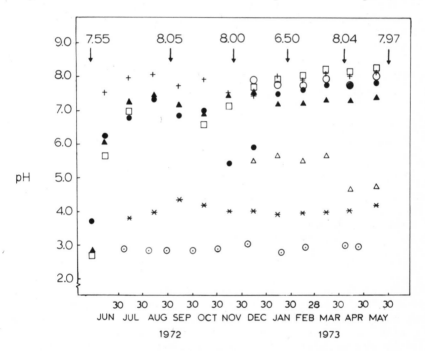

Figure 40. *Streams and rain water monthly pH averages as a function of time.*

Symbols used in Figures 40-45:
* *Rain water*
○● *Streams originating in experimental hill #1*
△▲ *Streams originating in experimental hill #2*
✚ *Streams originating partly from experimental hill #1 and partly from experimental hill #2*
□ *Streams originating from experimental hill #3*
⊙ *Streams originating from strip mined land adjacent to the experimental sites*
↓ *Streams originating in natural hardwood forest.*

The three streams that flowed from the experimental strip-mined land had a pH of 3.7, 2.9, and 2.7 when measured on May 31, 1972 prior to reclamation. These low pH values for stream waters are typical of the area and other streams originating from strip-mined land of similar geological overburden.[1,5] In comparison, the pH of stream waters of natural undisturbed land in the vicinity of the experimental hills fluctuated (as expected) between pH 6.5 and 8.1 depending on the time of the year.[10,19] The pH of the incoming rain water in this area varied within a narrow range between 3.6 and 4.2. The average for the year was recorded to be 4.0. The pH values of the incoming rain water and the stream water originating in the natural and strip-mined land indicates that acid formation in the unreclaimed land was substantial.

For comparison, total acidity of the three different water sources is reported in Table XXIII.

Table XXIII

Total Acidity (ppm CaCO$_3$) and pH of Various Water Sources
Guernsey County, Ohio

Sample	pH	Acidity
Rain	4.00	10.0
Natural stream	7.60	6.7
Affected stream	2.87	2200.0

Again these data confirm the pH measurements and strengthen the realization that acid formation through pyrite oxidation in the affected land was great. Acidity added to the land by rainfall (originating from air pollution) is very substantial when one deals with natural undisturbed land. It becomes minute and insignificant when compared to the total acidity of the affected strip-mined land.

After land reclamation, almost 6" of rain fell on the land during the 10-day period of Hurricane Agnes (June, 1972). The land was saturated with water and several streams sprang up in several different locations of the altered topography. The monthly pH values of the streams as a function of time are also shown in Figure 40. Immediately, the streams' pH rose to between 5 and 6. This was a sharp and unexpected rise attributed to the coal,

shale, and other potentially acid-forming strata
buried as deeply as possible in the overburden.
Throughout the experimental sites there was little
or no evidence of coal or coal-related strata at the
surface. This fact and spoil and soil water saturation substantially reduced acid formation. In all
sites, the streams' pH hyperbolically increased
during the summer to 7 and then decreased to pH 6
in all cases. This pattern followed the rainfall
frequency and intensity, which was high during June,
July, and early August, and low during the rest of
August, September, October, and part of November of
1972. The reduction in soil and spoil saturation
in early fall allowed for air introduction to the
mantle and subsequent acid formation.

In the fall, the pH level of the streams reached
a low point. However, from this low point the pH
level of the streams originating in experimental
hills #1 and 3 constantly increased until the pH
level of natural streams was reached. These results
were different from those obtained in the streams
originating in experimental hill #2. The pH level
of the stream was, in one case, at about 7 or below
and, at the other, it was about 5. The main difference between experimental hill #2 and the other two
hills is that top soil was not introduced to this
site.

In general, then, the acid formed in all three
experimental hills seems to have been reduced substantially when compared to streams originating from
other reclaimed land in the Fairview strip mining
field. Samplings of these streams revealed that the
pH of stream waters were below 3 even two years after
reclamation. Thus, it is clear that the careful
burial of coal, coal-related strata, shale, and
potentially acidic sandstone deep within the spoil
soon after the end of the strip mining activity may
have substantially reduced acid formation. The
addition of top soil as a dressing to the land further improved the inhibition of acid formation as it
was evident from the similar pH values of the streams
originating in the affected land compared to those of
streams originating in natural hardwood forest. In
addition, the lack of high walls, steep uncovered
slopes, and the complete plant cover achieved soon
after reclamation hindered the free flow of air that
is so essential to pyrite oxidation. The lengthening
of the path of air to the potentially acidic rocks
was highly successful in reducing acid formation
almost immediately after reclamation. Thus, all
water leaving the three experimental sites was of

acceptable quality and its effect on the biology and
ecology of the streams it flowed into was determined
to be minimal.

The oxidation of pyrite to sulfuric acid and iron
sulfate results in the increase in total acidity and
the reduction in stream pH. Another good barometer
of pyrite oxidation and extent is the presence of
sulfate in the streams. Data shown in Figure 41
presents the kinetics and dynamics of sulfate formation prior to and during one year of land reclamation.

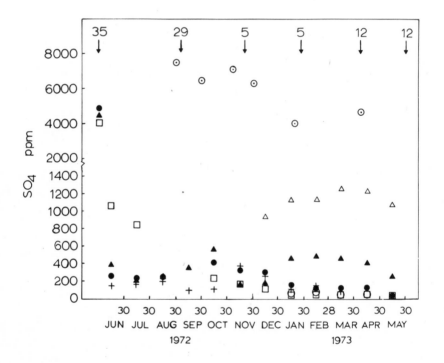

*Figure 41.   Streams monthly sulfate averages (ppm) as a function
of time.*

Generally, natural stream water contained only
minute levels of sulfate. It fluctuated between 5
and 35 ppm sulfate. Rain water sulfate content was
variable. In winter and spring it was low, while
during the summer it was as high as 50 ppm sulfate.

The streams of the strip-mined land contained above 4,000 ppm sulfate prior to reclamation. This acid condition was similar to sulfate contents of streams originating in other reclaimed land in the Fairview strip-mined fields. Thus, originally, the sulfate contents of the strip-mined land water was so very high as to indicate that we were dealing with an explosive acid condition. These results were similar to those reported for other strip-mined land streams.[1,5,7]

After reclamation, there was a sharp reduction (90%) in sulfate content of all the streams. This sharp reduction corresponded very closely with the reduction in stream pH and total acidity. During the year there were some fluctuations in sulfate content of the streams. In the October-November period there was a rise in sulfate content in all the streams. Thereafter, the sulfate content of all streams originating from experimental hills #1 and 3 gradually decreased to around 50 to 75 ppm, which is a level not harmful to any plant or animal living off the land or streams. The decrease in sulfate content of the affected streams was parallel to the decrease in their acidity. The streams originating in experimental hill 2 contained higher levels of sulfate, ranging between 400 and 1,000 ppm. These sulfate levels by themselves represent a major reduction in acid formation that took place in the spoil prior to reclamation (a reduction of at least 80% in acid formation). These are reported elsewhere in the literature.[1,5,7] Such sulfate levels cause only minor problems in the nutrition and development of higher plants.[20,21]

Sulfuric acid formation in the spoil banks further increases the solubility of many minerals found in the sedimentary rocks.[11] This increase in mineral solubility leads to an increase in total dissolved salts of stream water. The stream's dissolved salt load of our experimental field is presented in Figure 42 where the specific conductance (in mmhos/cm) is plotted as a function of time.

The specific conductance of natural stream water in the Fairview area ranged between 0.11 and 0.43 mmhos/cm throughout the year; that of the affected land prior to reclamation ranged between 1.75 and 3.3 mmhos/cm; that of other reclaimed strip-mined land between 2.7 and 6 mmhos/cm. From these figures it is apparent that the total dissolved salts of the affected land were substantially higher than for the case of natural stream water.[1,5,7,10,12] Such high specific conductance in water may hinder the survival

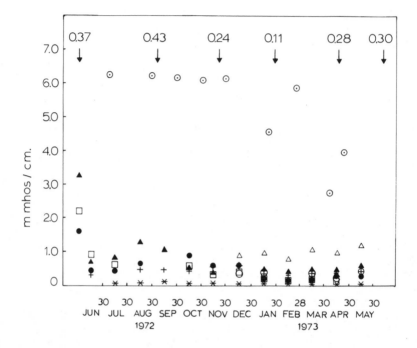

Figure 42. Streams monthly specific conductance (mmhos/cm) as a function of time.

and growth of some forms of life, but without the strict knowledge of the identity of the dissolved salts one would not be able to say what the effects would be on the stream.

After land reclamation, there was a steady decline in the specific conductance of the streams originating in experimental hills #1 and 3. One year after reclamation, total dissolved salts of the affected streams were only slightly higher than those of natural streams. In experimental hill #2 there was a noticeable reduction in specific conductance after reclamation (from 3.3 to about 1.0). However, there was little improvement thereafter. One year after reclamation, the streams of experimental hill #2 had a specific conductance of 0.5 and 1.0 mmhos/cm respectively. While there was no decrease in total

dissolved salts in these streams, there was also no
increase as a function of time. The total dissolved
salts reported for all streams including the streams
of experimental hill #2 by itself will not be considered very harmful to biological activity.[20,21]

The identity of the dissolved ions in the stream
water rather than the total dissolved salts is of
importance. Thus, a comprehensive survey of the
elemental composition of stream water was initiated
at the start of this field experiment.

The first obvious element to follow was iron
since it is the resultant product of pyrite oxidation.
Natural streams contain only minute quantities of
this element.[10] At no time during the year did the
natural stream water contain more than 0.1 ppm iron.
The streams' water of the affected land contained
between 25 and 75 ppm iron. This great variability
was noticed throughout our work. After reclamation,
the iron content of all stream water was always low—
below 1.0 ppm. These low iron levels are attributed
to its precipitation in water when the pH is above
3.5.[22] Since most of the streams originating in the
reclaimed strip-mined land had a pH of at least 5,
the iron content of the water was always very low
(below 0.1 ppm).

Manganese content of natural stream water is
generally low.[19] In the natural streams of the hardwood forests in the vicinity of the experimental
field, the manganese levels were found to be very
low (less than 0.1 ppm). In the streams originating
in the affected land prior to reclamation, the level
of manganese was recorded to be 80 ppm (Figure 43).
In other strip-mined land in the strip-mined fields
of Fairview, the manganese concentrations were between 100 and 300 ppm.[5,17] After reclamation, stream
water originating in experimental hills #1 and 3 had
manganese levels below 2.0 ppm. In most instances
Mn concentration was below 1 ppm. On the other hand,
the manganese levels of the streams originating in
experimental hill #2 were always between 2 and 10
ppm. However, even such concentrations were recently
demonstrated to cause a severe reduction of plant
growth, development and productivity.[23,24] Such
concentrations of manganese will only add very little
to the specific conductance of the water. However,
since this element is known to be essential for plant
growth in only very small amounts;[25] a small seem-
ingly insignificant increase in its concentration
leads to a drastic reduction in plant growth. The
removal of manganese from stream water is an absolute
necessity since it will stay in solution at low as

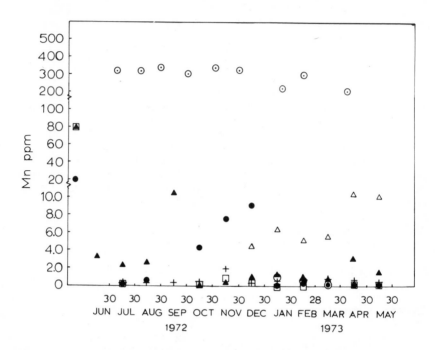

Figure 43.   Streams monthly manganese averages (ppm) as a
function of time.

well as high pH levels. Manganese will come out of
solution only at a basic pH of above 8.5.[22]
     Other heavy metals such as mercury, cadmium,
zinc, and copper were analyzed for but were found
to be in insignificantly low concentrations. For
that reason it was decided not to follow them.
     Two other cations, calcium and magnesium, were
found to be in high concentrations in the stream
water. Calcium and magnesium concentrations in the
various water sources are shown in Figures 44 and
45 respectively. From these results it is evident
that calcium as well as magnesium are found in
substantial concentrations in the streams of the
experimental field. The streams of the affected
strip-mined land contained amazingly high levels of
calcium and magnesium (400 and 250 ppm respectively).
Such concentrations of magnesium were shown in our
laboratory to be detrimental to plant growth.[26]
Upon reclamation, the levels of calcium were reduced

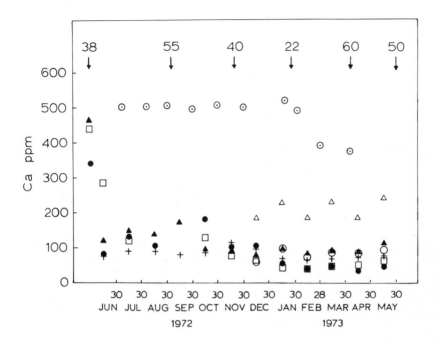

Figure 44.   Streams monthly calcium averages (ppm) as a function of time.

to about 100 ppm in all streams and the level of
magnesium in streams originating in experimental
hills #1 and 3 steadily decreased to the levels of
natural streams.[19] However, the calcium and magnesium
levels remained at substantially higher levels (150
and 40 ppm respectively) in experimental hill #2.

   The predominance of calcium in stream water is
of prime importance since this element regulates the
transfer of nutrients across plant membranes by
selectively acquiring certain elements and rejecting
others.[27,28]   Thus, the higher the calcium levels
found in the streams, the higher the chances are
that nutrients detrimental to life will not cross
the biological membranes and cause much harm.[23,24,27]
The calcium levels registered in the streams after
land reclamation in experimental hills #1 and 3 are
within this range.

   Magnesium is another element essential to life.[25]
It has little to do, however, in regulating the uptake

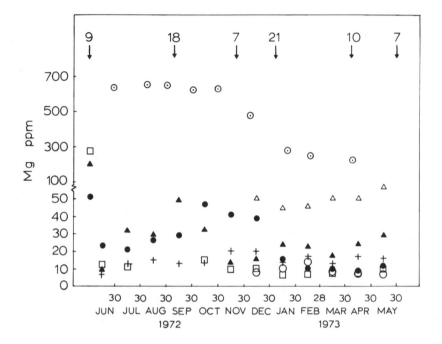

Figure 45. Streams monthly magnesium averages (ppm) as a function of time.

of elements across plant membranes as calcium does. It is needed by the cell in large quantities; however, when it is found in such concentrations as in streams originating in partially reclaimed land, it can lead to serious reduction in plant growth and cell development.[20,21] However, when its concentration is reduced to the levels reported for all streams originating in the experimental hills, it can provide a benefit to the living cells. Basically, in concentrations of 50-100 ppm it can severely reduce manganese uptake by plant cells.[23,24] This reduction in manganese uptake seems to be a competitive inhibition.

The levels of sodium and potassium in the stream water were low and they were regarded to have little or no significant effect on the ecology or biology of the system.

The water chemistry results discussed in the preceding paragraphs indicate that severe acid conditions existed in the streams of all experimental hills prior to reclamation. Following reclamation, the stream quality improved in experimental hills #1 and 3 to an extent that reached a stage similar to natural stream water. In experimental hill #2 there was a definite improvement in the water quality; however, some acid formation continued to take place on this hill. Even so, the quality of stream water originating in experimental hill #2 was substantially better than for other reclaimed lands in the Fairview strip-mined fields.[1,5]

Other data collected show some additional differences among the experimental hills. The data are briefly described below:

1. The conditions existing in experimental hill #3 were such that erosion was very low because of land terracing. The streams did not carry over 500 ppm of suspended solid matter. The same holds true for experimental hill #2. The situation in experimental hill #1 was drastically different. Immediately after reclamation, erosion was great, amounting to 30,000 ppm suspended solid matter in the streams during heavy rains. The erosion originating from experimental hill #1 resulted in great loss of the expensive top soil introduced to the land. Top soil of experimental hill #1 was lost out of the system while in the other two hills it was prevented from leaving the land by the contours and terraces. Thus, it is highly recommended that the strip-mined land be terraced to stop land erosion.[1,6,7]

2. Land revegetation was good to excellent in the three experimental hills. In hill #3 plant growth was heavy. During the spring of 1973, the land was covered with a healthy growth of legumes and grasses. The chemical composition of the plants was similar to those grown under normal conditions. The heavy and healthy plant growth of experimental hill #3 was due mainly to:

    a. high amount of water available in the land because of the extensive terracing
    b. high amount of water held by the soil introduced to the land
    c. high soil pH (6.0) indicating low acid formation and thus low levels of toxic elements in the soil solution.

These favorable conditions insured good plant growth throughout the year.

In experimental hill #1 there was initially a
good plant cover. However, because of the high rate
of erosion and the lack of terracing to stop loss of
water from the land, the amount of water available
for plant growth was low. Through the summer and
fall months the soil was dry and plant growth stopped.
The low soil moisture content, in addition to the
constant exposure of deeper spoil material through
land erosion and gully formation, allowed for air
penetration into the spoil and thus resulted in some
acid formation. This acid formation was evident in
the low soil pH (4.0) and high sulfate (400 ppm) and
other soluble salts found in soil solution. Thus,
as a result, experimental hill #1 did not support as
good a plant cover as did hill #3. It also is
anticipated that increase in acid formation will
essentially result in reduced stream quality as land
erosion continues.

Soil was not introduced to experimental hill #2.
Thus, seed germination was the poorest of the three
treatments. However, because of the terracing there
was a substantial plant growth above, on, and below
the terraces. There was an apparent acid formation
on this spoil because the spoil pH was 4.0 and the
sulfate content (500 ppm or above) and other soluble
salts were significantly higher than in experimental
hill #3.

Soil addition to terraced land is highly recommended. Terracing without addition of soil will not
result in good plant growth and ultimate stoppage of
acid formation. On the other hand, addition of top
soil to nonterraced land will not result in good
plant cover nor would it completely stop acid
formation.

3. Chemical plant analysis of vegetation
collected from the three experimental hills indicate
that they were not significantly different from each
other or from vegetation grown on nonaffected land.

4. Water quality of springs, wells, and
ponds collected in the vicinity of the experimental
fields were not significantly different from those
present in natural surroundings.

In conclusion, this work shows that great care
exercised in burial of coal, coal-related strata,
and other potentially acid-forming rocks in addition
to land-terracing, and top soil application will
result in fast land revegetation, substantial reduction in land erosion, and reduction in acid mine

drainage. It was shown that the water quality of streams originating from such land is good enough to support plant and animal life. Farms in the vicinity of the experimental field had no complaints concerning water quality.

REFERENCES

1. Elzam, O. E. "An Ecological-Economic Study of Strip Mining in Southeast Ohio," Case-Western Reserve University Publication (1973).
2. United Nations. "Statistical Yearbook, 1971-1972," New York (1972).
3. U.S. Department of Interior. "Surface Mining and the Environment," U.S. Govt. Printing Office, Washington, D.C. (1970).
4. U.S. Department of Interior. "Mineral Yearbook, 1970," U.S. Govt. Printing Office, Washington, D.C. (1972).
5. Elzam, O. E. "Strip Mining and Reclamation: A Systematic Study of Four Ecosystems in Two Ohio Regions," Case-Western Reserve University Publication, National Science Foundation Report (1973).
6. Stack, J. F. "Strip Mining," Sierra Club Publication, New York (1972).
7. U.S. Geological Survey. "Hydrologic Influence of Strip Mining," U.S.G.S. Professional Paper 380 (1963-70).
8. Consolidation Coal Company Central Division, Cadiz, Ohio, personal communication.
9. Zanesville Flight Service Station, Zanesville, Ohio, personal communication.
10. Elzam, O. E. "Data Available from Experimental Station Near Fairview, Ohio," directed by the author, on behalf of Consolidation Coal Co., Central Division, Cadiz, Ohio (1973).
11. Longwell, C. *Physical Geology*. (New York: John Wiley & Sons, 1969).
12. Likens, G. E., F. H. Borman, N. M. Johnson, and R. S. Pierce. "The Calcium, Magnesium, Potassium, and Sodium Budgets for a Small Forested Ecosystem," *Ecology, 48*, 772 (1967).
13. Environmental Protection Agency - U.S. EPA. *Methods for Chemical Analysis of Water and Wastes* (1971).
14. Chapman, H. D. and P. F. Pratt. "Methods of Analysis for Soils, Plants, and Waters," University of California publication (1961).
15. Elzam, O. E. and E. Epstein. "Salt Relations of Two Grass Species Differing in Salt Tolerance. I. Growth and Salt Content at Different Salt Concentrations," *Agrochimica, XIII,* 187 (1969).

16. Johnson, C. M. and A. Ulrich. "Analytical Methods," California Agr. Exp. Station Bulletin #766 (1959).

17. Elzam, O. E., *et al*. "The Effect of Strip Mining on a Natural System: A Water Quality Study of Piedmont Lake, Ohio," Case-Western Reserve University Report (1970).

18. U.S. Department of Interior. "Stream Pollution by Coal Mine Drainage Captina Creek Basin, Ohio," Work Document #23 (1968).

19. Elzam, O. E. "Hydrogen, Sodium, Potassium, Calcium, and Magnesium Stream Chemistry of a Small Forested Grassland Ecosystem in Northeast Ohio," Proceedings of the 16th Conference on Great Lakes Research (1973), pp. 140-141.

20. Berstein, L. and H. E. Hayward. "Physiology of Salt Tolerance," *Ann. Rev. Plant Physiol.*, *9*, 25 (1958).

21. Richards, L. A., Ed. *Diagnosis and Improvement of Saline Soils*. U.S. Dept. of Agriculture Handbook 60. (1964).

22. Hill, R. D. and R. C. Wilmoth. "Limestone Treatment and Acid Mine Drainage," Proceedings SME (1970).

23. Elzam, O. E. "Reversal of Manganese Plant Toxicity with Calcium and Magnesium Addition to Strip Mined Land," Case-Western Reserve University Report (1973).

24. Katz, L., R. Bartolatta, M. A. Topa, and O. E. Elzam. "Manganese Relations of Barley Plants," Proceedings Midwest Annual Meeting of the Society of Plant Physiologists (1973).

25. Johnson, C. M., P. R. Stour, T. C. Broyer, and A. B. Carlton. "Comparative Chlorine Requirements of Different Plant Species," *Plant and Soil*, *8*, 332 (1957).

26. Elzam, O. E. "Laboratory Demonstration Experiments," Case-Western Reserve University (1973).

27. Epstein, E. "The Essential Role of Calcium in Selective Cation Transport by Plant Cells," *Plant Physiol.*, *36*, 437 (1961).

28. Rojas, E. and J. M. Tobias. "Membrane Model: Association of Inorganic Cations with Phospholipid Monolayers," *Biochim. Biophysi. Acta*, *94*, 394 (1965).

SECTION V

SOCIAL FACTORS

19. INTRODUCTION

In the preceding sections of the book we have
considered the availability of mineral resources
including energy and water, the pollution problems
involved in resource extraction, and the technical
problems of reclaiming or disposing of discarded or
displaced resources. Other areas that will now be
considered include both the social and economic
impact that extraction of resources has had on the
development of our civilization.

Our modern society is very heavily dependent on
mineral resources. A sizeable portion of our gross
national product (GNP) represents the efforts of the
mineral industry. Much of our affluence depends on
the continued extraction of minerals that are used
to fabricate the appliances that work for us, the
cars and planes that move us, and the homes and
buildings that house us. Without minerals we would
have to return to the time of the cavemen.

But minerals not only account for our higher
standard of living, they are also responsible for
changes in our living style and an increase in waste
discards. Our resource-based society is a mechanized
one. Machines built using minerals can do the work
formerly done by many men. Some machines designed
by man are even able to conduct logical thought processes at speeds many times faster than humans.
This mechanization has moved society from rural to
urban areas and more recently from urban areas to
suburban developments. Such developments are tending
more and more to becoming small cities with all the
services and shops needed for the community. In some
areas the suburban sprawl has even brought the down-
fall of the older central city. Movement of people
out from the central city usually involves the most
affluent, leaving behind the poor and the minorities.
This flow has left many of our large cities with a
much reduced tax base and has forced them to curtail
some basic services for lack of funds.

233

Our affluence has also led us to become more
wasteful. New ways of packaging, nonreturnable bot-
tles, canned products, plastic containers, and many
other conveniences of our time not only add ease to
our living but also add a tremendous cost to our
government units who are responsible for their ultimate disposal. A point will be reached when the
economic benefit gained by the use of "disposables"
will be less than the cost of their disposal. Some
say we are already past this point. Others go further
and say that environmental considerations should take
priority even if the economics are unfavorable.

It all comes down to the idea of trade-offs.
Yes, resource extraction damages the environment—
but we need resources in our society. A balance must
be achieved by proper planning to reduce negative
effects while maximizing efficiency of resource
utilization. Economists as well as environmentalists
normally agree on this point.

Ray, in the first paper of this section, discusses the very important subject of measured human
attitudes versus stated opinions regarding the extraction of mineral resources. His paper addresses
itself indirectly to the problem of trade-offs that
we have just discussed. His conclusion, that measured
attitudes show strong opposition toward surface mining
while an equal number of respondents stated negative
and positive opinions toward surface mining, shows
our lack of truthfulness in discussing this issue
and the difficulty of arriving at a rational policy
decision that is in the best interest of society.

Treacy and Overton further examine strip mining
and consider a way to measure the economic advantage
or disadvantage that stripping poses on a given area.
How does mining affect property values?

The final paper in this section, written by Clark
and Goddard, summarizes the overall changes in our
living style, the trends toward suburbanization and
decentralization, and the solid waste disposal crisis.

20. STRIP MINING IN OHIO: A COMPARISON OF
MEASURED HUMAN ATTITUDES AND STATED OPINIONS

John R. Ray
Wright State University
Dayton, Ohio

ABSTRACT

This study investigates the nature of that portion of
the social milieu reflected in human attitudes and stated
opinions of a sample population in Ohio toward the concept of strip mining for coal. Data were collected with
a questionnaire and an attitude scale constructed for the
project, using Thurstone's method of paired comparisons.

The study area included 11 counties selected to produce
a spatial bias in the data. A random selection of approximately one-third of the townships in each county was
made, resulting in 54 townships being identified as
data-collecting areas.

An analysis of measured attitudes revealed an overwhelming negative reaction by the sample population
toward surface mining. Only 15% of the population had
a positive attitude toward the concept, while 5% established a neutral position on the scale.

However, the population was divided with respect to
stated opinions on the concept. Thirty-eight per cent
of the sample had positive opinions toward stripping;
an equal proportion had negative opinions on the subject.
Twenty-four per cent refused to state an opinion and
elected the neutral shelter provided in the questionnaire.
The majority of persons expressing a positive opinion
toward stripping revealed a recognition of the need for
a source of fuel and power. Those expressing a negative
opinion were concerned about the destruction of the land
and local communities by surface mining operations.

Strip mining began in Ohio in 1914. This activity was initiated by the sudden increase in demand for coal caused primarily by the advent of World War I. From 1914 to 1948 the industry operated essentially without control or regulation. This resulted in the serious alteration of 45,213 acres of land in 25 counties in southeastern Ohio[1] (Figure 46). Very little effort was made during this period

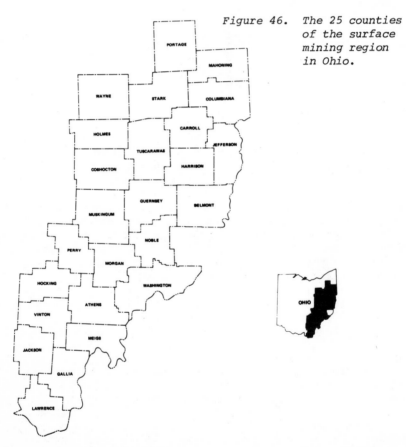

Figure 46. *The 25 counties of the surface mining region in Ohio.*

to restore these lands. There were a few concerned individuals and coal operators who did some seeding and planting of trees on affected lands during this period, and in 1941 a small group of operators formed the Ohio Reclamation Committee. This marked the beginning of a concerted reclamation effort in the state. In 1945, the Ohio Reclamation Association was organized. This association is a private,

voluntary organization of surface mining operators
charged principally with research in methods of
reclamation of strip-mined lands.[2] By 1948 the
association and its members had, among their several
actions, managed to assist in the planting of trees
on 3,764 acres of stripped lands.

During this initial period private citizens and
special interest groups became concerned over the
environmental alteration caused by strip mining.
Their concern was demonstrated by sporadic efforts
to cause regulation of strip mining by legislation.
Although these individuals and groups had interests
that were often divergent, their activities in pur-
suit of their perceived common enemy, strip mining,
were instrumental in providing the necessary support
for passage of Ohio's first strip mine legislation
in 1947. This legislation became effective in 1948.
It has been this active segment of the population of
Ohio that has borne primary responsibility for the
modifications and level of implementation of strip
mining regulatory legislation since that time.

In the last 25 years the industry has had to
function in what has amounted to a changing social
environment. Periodic reaction by the concerned
individuals and special interest groups in Ohio's
population to industry efforts to respond to increasing demands for coal has been effective in
causing the original legislation to be amended and
modified at least five times. The first amendment
was accomplished in 1949 when the legislature accepted recommendations from the industry for the
licensing of operators, surface grading, and planting
of affected areas. In 1955, there was an operational
up-dating of the law, and in 1959 the state was
authorized to acquire and develop unreclaimed strip-
mined land. In 1965, there was another operational
up-dating of the law with major revisions in the
reclamation provisions. More recently in 1972, a
substantial revision of the law was accomplished,
making Ohio's strip legislation perhaps the most
comprehensive control law in the nation.

These modifications document the generally successful campaigns of a select group of persons within
the state to cause the regulation of an industry
which by its very existence is either preventing or
discouraging a variety of activities by these groups
or individuals, or is providing them with an issue to
exploit. However, there has been no documentation of
the nature of the social environment in Ohio as it
relates to the interests and desires of the whole
population and the extent to which these are in

conflict or agreement with the activities of the
strip mining industry. Specifically, no research
has been completed that reveals just how the population of Ohio perceives the particular aspects of its
environment affected by the surface mining industry's
attempts to meet increasing demands for coal.

It was the object of the research reported here
to examine the attitudes of the population of Ohio
to determine if there is general support or agreement
with the effort of the strip mining industry to produce coal in Ohio. The effort has involved the
measurement of human attitudes and collection of
stated opinions of a sample population in Ohio as
they relate to the concept of strip mining for coal.
Such information should be useful in a number of ways
to the industry, government agencies, public and
quasi-public interest groups, and concerned individuals. With the establishment of a measure of the
nature of this characteristic of the population it
may be possible to identify some of the factors in-
fluencing the development and modification of this
aspect of the social environment. Such knowledge
could be very useful to those who are stimulated or
required to respond to the continuing efforts of the
industry to meet the demand for coal.

Specific tasks were required before data could
be collected and analyzed to accomplish the basic
objective. First, a study area had to be defined;
second, a questionnaire had to be prepared; and
third, an attitude scale had to be constructed.

The study area selected includes 11 counties
(Figure 47). Four—Butler, Greene, Pickaway, and
Perry counties—are about equally spaced across the
southern part of the state, while four others—
Henry, Wyandot, Richland, and Coshocton—are about
equally spaced across the northern part. These eight
counties extend westward from a *Core Area* of strip
mining in eastern Ohio. From these eleven counties
a random selection of approximately one-third of the
townships in each county was made. This procedure
provided 54 data collecting areas from which 432
interviews were completed. The exact number of
interviews completed in each county was influenced
by the total number of townships selected in the
county.

The questionnaire was designed to elicit data on
certain economic, personal, and social variables from
the study population. An analysis of several of
these variables revealed that the population possessed
a rather broad range of economic and social characteristics. More precisely, it was impossible to

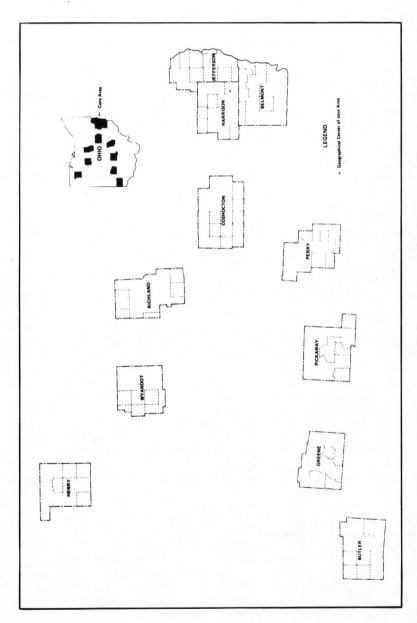

Figure 47. Study area: Ohio counties and townships selected for sampling attitudes toward strip mining.

characterize the sample with respect to these several variables. Such a sample was then considered to be representative of the total population of the state. An additional feature of the questionnaire was the opportunity it provided for respondents to make a verbal statement regarding their opinion on strip mining and the opportunity to state a reason for the opinion expressed.

The attitude scale was constructed following Thurstone's Method of Paired Comparisons.[3] In constructing the scale a series of 30 statements ranging from extremely positive to extremely negative meanings regarding strip mining were extracted from several sources. These statements were then modified in an effort to meet most of the requirements established for scale stimuli.[3] Fifteen persons who had previously indicated either positive or negative opinions toward strip mining were given the task of ranking these statements following their personal reaction to the subject involved. Then a set of fifteen statements was selected from these rankings, representing those most frequently selected at positions along the continuum of rankings by the fifteen persons.

These fifteen statements were interpaired and the resulting 105 pairs of statements were subjected to comparative judgments by a population of 100 persons. Each was asked to indicate which stimulus in each pair was preferred or reflected their attitude toward strip mining. Seventy sets of responses were satisfactorily completed. These were used to complete the scale construction. An empirical frequency (F) matrix was computed from these comparative judgments. The frequencies corresponded to the number of times each statement was judged more favorable than the one with which it was paired. The frequency matrix was converted to a proportion (P) matrix, and using tables available, the cells of the (P) matrix were converted to unit normal deviates, producing a normal deviate (Z) matrix.

Each column of the (Z) matrix was summed and the mean value of each of the fifteen columns, representing the fifteen statements, was computed. The column mean values are accepted as the scale values along a psychological continuum for the fifteen stimuli. The procedure described for computing the scale values has been demonstrated to be a least-squares solution for the model.[4]

From the fifteen stimuli scaled, seven were selected to provide approximately equal intervals on the continuum. This reduction in the number of

stimuli was made to provide a more reasonable device
for field use in attitude measurement. The seven
stimuli were then subjected to a chi-square test to
allow an examination of the level of difference
between the expected and empirical proportions. The
chi-square test is based on an *inverse sine trans-
formation* of the expected and empirical proportions.[5]
In essence, the test provides a means of determining
whether certain assumptions involved in construction
of the scale, were tenable. The most important of
these assumptions is the unidimensionality of the
psychological continuum, *i.e.*, that the distances
marked off by scale values along the continuum are
additive. For the seven stimuli a chi-square =
16.16 was computed. This value is not significant
at the 0.01 level. Thus, the assumptions involved
in constructing the attitude scale were considered
tenable, and the scale was assumed to be adequate
for use in this study. The scale is presented in
Table XXIV with the stimuli arranged in numerical
sequence based on the positions of the stimuli on
the continuum. In the field, the scale stimuli were
arrayed in a random order.

Table XXIV

*Paired Comparison Attitude Scale Values and Stimuli
for the Concept: Strip Mining for Coal*

Scale Values	Stimuli
0.000	Strip mining provides benefits for individuals and the community for which the coal companies are not rewarded.
0.216	We should reward and acclaim those who are capable of extracting a profit from coal lying beneath the surface.
0.434	Strip mining should be halted because it produces serious damage to a familiar landscape.
0.637	Strip mining for coal is properly described as a "rape of the land."
0.935	One should not destroy what he cannot recreate.
1.045	One should be very concerned about the amount of land taken out of agricultural and recreational use by strip mining.
1.278	Any judgment of the social value of strip mining should be a matter of balanced thinking.

     In the interviews each respondent was asked to
mark three of the seven stimuli with which he/she
could agree, or which most nearly represented his/
her attitude toward strip mining.   The median scale
value for the three stimuli was used as the scale
position for each respondent.[3]   Thus, a respondent
marking statements having scale values of 0.434,
0.935, and 1.278 was scaled at 0.935 and was assumed
to have a negative attitude toward the concept scaled.
     There are several factors inherent in this method
that must be kept in mind when undertaking analysis
of the data gathered by this technique.   Among these
is the question of the influence of the bias of a
single group of judges on the scale positions of a
set of stimuli.   Another is the question of whether
a different group of judges would produce the same
or a different set of values for the identical set
of stimuli.   There is also the question of whether
a particular set of statements when used to measure
attitudes of a population will elicit responses
indicative of the attitudes of a population with a
variety of experiences and levels of knowledge about
the concepts being studied.
     With reference to the first question, it should
be clear that it is desirable to use as judges
people who are from a population similar to the one
from which the subjects included in the research
sample will be selected and to whom the finished
scale will be applied.   Wright State University
students were used as judges in the construction of
the scales used in this study.   In part, the students
were used as judges in order to expedite the task of
scale construction.   However, the particular groups
of students selected possessed characteristics that
made them acceptable as judges in this study.   They
were from a student body that commuted to the university from a fifty-mile radius and represented a
variety of social and economic levels.
     With reference to the second question, investi-
gations have shown that scale values for the same
stimuli obtained from widely differing groups of
judges correlate rather highly with one another.[3]
Inevitably, all judges will be influenced by their
own attitudes, but these investigations suggest that
in most cases the effect will be small and have a
minimal effect upon the order of scale stimuli.
     In response to the third question, the statements
selected for judging were carefully chosen and edited
in accordance with prescribed criteria.   This does
not absolutely insure that each of the statements
included will differentiate accurately between those

subjects with favorable and those with unfavorable attitudes. As in all attempts to measure attitudes by means of scales, the subject's position on the attitude continuum is unknown and must be estimated from his responses to the stimuli contained in the scale. Only extensive use and testing of the statements contained in the scale will demonstrate their ability to discriminate effectively. This is true, in spite of the confirmation of their suitability for use in this study by the chi-square test.

Finally, it must be recognized that any given set of stimuli represents only a small number of statements from an infinitely large universe of statements about a concept. As such, the set may fail to include any statements reflecting the particular feeling or belief of a subject who is asked to select three of the seven statements which most closely reflect his attitude toward the scaled concept. It should be remembered, however, that the stimuli have scale positions that were previously established by a group of judges. Thus, the process of identifying the three most acceptable statements of the seven presented for the concept will allow the attitude of the subject to be estimated with respect to its direction and relative strength.

The attitude scores on the concept, strip mining for coal, for the study population are presented in Table XXV along with an interpretation on a verbal scale. The results suggest very strongly that attitudes toward strip mining in the study population are negative. There were 347 persons, or 80.0% of the total population, who revealed negative attitudes toward this concept on the paired comparison scale. There were 22 persons, or 5.0% of the population who revealed neutral attitudes, and 63 persons, or 15.0% of the sample who revealed positive attitudes toward strip mining. The scale provides a measure of the strength of these attitudes as well. In column four of Table XXV there are 181 very negative attitudes and 166 negative attitudes recorded. At the opposite end of the scale, 57 very positive attitudes and 6 positive attitudes are shown.

In comparison, Table XXVI presents data for an analysis of the verbal opinions of the respondents on strip mining. An examination of these data reveals a striking change in the reaction of the sample population toward the concept: 163 persons, or 38.0% of the group gave negative opinions, while 104 persons, or 24.0% of them, gave neutral opinions; 165 persons, or another 38.0% of the group gave positive opinions. Thus, the sample population had about an

Table XXV

Paired Comparison Measure of Attitudes Toward Strip Mining for Coal

Median of Scale Scores	Scale Score Interpretation	Number of Respondents	Grouped Interpretation	Number of Respondents	Per Cent of Total
0.216	Very Positive	57			
0.434	Positive	6	Positive	63	15
0.637	Neutral	22	Neutral	22	5
0.935	Negative	166			
1.045	Very Negative	181	Negative	347	80
	Total	432		432	100

Table XXVI

Stated Opinion on Strip Mining for Coal

Stated Opinion	Number of Respondents	Per Cent of Total
Positive	163	38
Neutral	104	24
Negative	165	38
Total	432	100

equal number of persons with divergent opinions on strip mining, while a significant number of persons elected to refrain from stating an opinion on the concept.

Table XXVII presents an array of the reasons given for the appeal of strip mining to those respondents expressing a positive opinion on the concept. The statements represent a synthesis of the great variety of ways in which respondents reacted toward this entry in the questionnaire. A synthesis of these stated reasons for the appeal of strip mining was made to provide a minimum number of statements among which distinction could readily be made for analytic purposes, but which would as accurately as possible reflect the expression given by the individual respondents.

From the data provided in Table XXVII the study population revealed a high level of recognition for the need of a source of power and fuel. This is indicated by the number of responses identified with statement number three. Some interest was expressed in the role strip mining plays in providing employment, locally. This is indicated by 34 responses that can be identified with statement number two.

Table XXVIII provides an array of reasons given for objecting to strip mining for coal by those persons expressing a negative opinion on this concept. Again, these statements represent a synthesis of the variety of ways in which respondents reacted toward this entry in the questionnaire.

From the data provided in Table XXVIII it appears that those persons expressing a negative opinion on strip mining seem most concerned with the general deterioration of the land, the appearance of an area

Table XXVII

Reason for Appeal of Strip Mining for Coal Given by
Respondents Expressing a Positive Opinion on the Concept

Stated Reason for Appeal of Strip Mining	Number of Respondents	Per Cent of Total
1. Represents a wise use of a natural resource	1	0.2
2. Provides employment for the local population	34	7.9
3. Satisfies the demand for power and fuel	109	25.2
4. Provides revenue for the state	1	0.2
5. Provides numerous benefits for the community	5	1.2
6. The safest method for mining coal	6	1.4
7. Provides income for land owners and coal companies	3	0.7
8. Represents the best use for most of the land which is stripped	2	0.4
9. Represents an efficient, low-cost method of meeting the demand for coal	2	0.4
Total	163	37.6

following the stripping, and the degrading of other resources associated with the affected land. This is evidenced by the number of responses that can be identified with statements one, two, and four. Some concern was expressed over the low rate of restoration perceived by some and environmental pollution and waste of land perceived by others. This concern is evidenced by responses that can be identified with statements three, seven, and eight.

There is, then, a considerable difference between the opinions expressed by the study population toward strip mining and their measured attitudes. Whereas, 63 persons, or 15.0% of the population of 432 persons, revealed a positive feeling toward strip mining as measured by the attitude scale, 163 persons, or 38.0% of the group, expressed a positive opinion toward strip mining when given the opportunity to make a verbal response to the concept. Twenty-two

Table XXVIII

Reason for Objection to Strip Mining for Coal Given by
Respondents Expressing a Negative Opinion on the Concept

Stated Reason for Objection to Strip Mining	Number of Respondents	Per Cent of Total
1. Destroys the natural beauty of the land	33	7.6
2. Leaves the land in a poor condition for any use	32	7.4
3. Is rarely followed by restoration of the land for future use	16	3.7
4. Causes total destruction of the land—soil, water, vegetation	48	11.1
5. Creates safety hazards for people and wildlife	6	1.4
6. Leaves the land in a useless condition, reducing the local tax base	1	0.2
7. It is a noisy, dusty, and dirty activity	10	2.3
8. It is a wasteful use of the land resource	14	3.2
9. It destroys total communities— houses, farms, churches, neighbors	5	1.2
Total	165	38.1

persons, or 5.0% of the population, had attitudes estimated as neutral on the paired comparisons scale; yet 104 persons, or 24.0% of the group made neutral responses toward the concept when they were provided an opportunity to respond verbally.

These contrasting sets of responses to the concept strip mining for coal have been obtained from two quite different instruments. The attitude scale consists of a series of belief and behavioral intention statements. Respondents were confronted with this series of statements and on the basis of their reaction to them an estimate of their attitude was obtained. The attempt here was to arrive at a single numerical value that would indicate how favorable or unfavorable the individual's attitude was toward the object or concept. This approach to attitude

determination is useful in many ways for it provides
a quantitative indicator of attitudes. Comparisons
of attitudes with a variety of social, economic, and
physical variables is possible when such information
on attitudes is provided. Further, the manner in
which the attitude scale is constructed and adminis-
tered provides a level of control over the situation
surrounding respondents at the moment attitudes are
being measured. Thus, a number of biases that could
influence the reaction of respondents are reduced or
eliminated by the use of an attitude scale.

In contrast, the interview technique employing
a questionnaire elicits a behavioral reaction by the
respondent toward the concept under study. The
technique allows only the range and distribution of
behavior toward the concept within the population to
be measured.[6] Further, control mechanisms are difficult to employ in the interview situation, and
biases tend to influence the behavior of respondents.
Specifically, the respondent may be influenced by
the personality of the interviewer, the stated purpose of his or her presence, their organizational
association, or any number of other stimuli that may
exist at the moment of the interview. The interview
technique also allows the respondent to easily effect
any type of trade-off that seems desirable to him
at the moment he or she is asked for a verbal commitment regarding the concept presented.

What is provided by the interview technique is
the opportunity to gather some indication of the
factors considered important to the respondent with
respect to his or her reaction to the concept under
study.

The data assembled in this study from the attitude scale provided an established pattern of attitudes in the population toward strip mining for coal.
That is, there is a learned predisposition to respond
toward the concept in a consistently negative manner
by the majority of the population.[7] This *feeling*
toward strip mining is very strong in a significantly
large proportion of the population. However, this
predisposition to respond is not consistent with
observed behavioral patterns toward the same concept.
Specifically, 38.0% of the population expressed a
high level of awareness of positive factors associated
with strip mining, *i.e.*, the need for energy, and
chose to take a positive position in support of strip
mining. However, an equally large proportion of the
population behaved in a manner consistent with their
measured feeling toward strip mining by taking a
negative position when asked for their opinion toward
the concept.

It is assumed that each person in the study
population was about equally aware of the aspects of
strip mining considered valid reasons for the existence of the industry by many and equally aware of
the reasons considered sufficient for its abolition
by others. So, among the persons in two of the three
groups identified, there was a willingness to effect
a trade-off between the desirable and undesirable
aspects of surface mining when they found themselves
in a situation requiring a behavioral response.
Finally, a third group in the population, 24.0% of
the total, when asked to react behaviorally to the
concept refused to allow a trade-off and chose the
neutral shelter provided.

REFERENCES

1. Department of Natural Resources, Division of Forestry and
    Reclamation. "Coal Strip Mining Statistical Report by
    Counties to January, 1969," Columbus, Ohio (1970).
2. Ohio Reclamation Association. "Tree Planting Guide for
    the Reclamation of Strip Mine Lands in Ohio," Technical
    Bulletin No. 70-1, Columbus, Ohio (1970).
3. Edwards, A. L. *Techniques of Attitude Scale Construction.*
    (New York: Appleton-Century Crofts, 1957).
4. Mosteller, F. "Remarks on the Method of Paired Compari-
    sons: I. The Least Squares Solution Assuming Equal
    Standard Deviations and Equal Correlations," *Psychometricka,*
    XVI, 3 (1951).
5. Mosteller, F. "Remarks on the Method of Paired Compari-
    sons: II. A Test of Significance for Paired Comparisons
    when Equal Standard Deviations and Equal Correlations are
    Assumed," *Psychometricka, XVI,* 207 (1951).
6. Allport, G. W. "Attitudes," in *Readings in Attitude
    Theory and Measurement,* M. Fishbein, Ed. (New York:
    John Wiley & Sons, Inc., 1967).
7. Fishbein, M. "Attitude and the Prediction of Behavior,"
    in *Readings in Attitude Theory and Measurement,* M. Fishbein,
    Ed. (New York: John Wiley & Sons, Inc., 1967).

21. MEASURING EXTERNALITIES OF STRIP COAL MINING VIA PROPERTY TAX ASSESSMENTS

John Treacy
Wright State University
Dayton, Ohio

Rosilyn Overton
Wright State University
Dayton, Ohio

ABSTRACT

The difficulty in measuring the costs to society resulting from strip mining has long been recognized. This
paper suggests that a medium for measuring in dollars and
cents, the effect of strip mining on a locality exists in
the tax valuation assessment process. A simple, easily
understood mathematical method of measuring external
costs is developed and tested on data from counties in
the state of Ohio.

INTRODUCTION

Growing concern over the quality of the environment has caused both federal and state governments
to enact legislation dealing with strip mining.
Both the operating methods and the reclamation of
the land after mining have been regulated.

From an economist's point of view, this legislation could be seen as an effort to correct the im-
perfect workings of the market mechanism in a
situation where there exist significant external
costs, or negative neighborhood effects. These
external costs or losses occur because the effects

251

of the transactions between coal producers and coal users extend to third parties who otherwise play no part in the transactions.

This is a case where the true costs of the good to society is not reflected in the price the buyer pays, nor in the cost figures of the producers. Whenever the market price of a good does not reflect its true cost to society, an inefficient resource allocation pattern will develop. Both producers and consumers will ignore the outside consequences and external costs in their decision-making. Consequently, they will tend to consume and produce more of that particular good than is socially desirable.

Thus it is seen that legislation requiring the adoption of certain types of production techniques and reclamation efforts can be interpreted as attempts to force the producers to internalize all costs, causing more efficient resource allocation. The resulting price structure from the revised techniques would cause buyers to consider the true cost when making decisions on quantity and alternate products.

Policy makers are faced with the difficulty of identifying the sources of externalities and then with developing adequate instruments for shifting the externalities into internal costs for the producers, without imposing costs in the process that are higher than the social benefit gained.

In the strip mining industry, the task of policy makers is further complicated by two factors: (1) the difficulty of understanding the dynamic effects of time on external costs of strip mining land, and (2) complaints that uneven-handed attempts to force internalization of costs will give a competitive advantage to those producers escaping the effects of regulation. If one segment of the energy industry is regulated with regard to external costs when others are not, the undesirable social costs of pollution will not be avoided, since buyers will tend to go to the lowest priced source of energy.

In the meantime, producers facing rising costs that put them at a competitive disadvantage will argue that the negative effect upon employment in the region in which the producer is located should also be considered by the policy makers. In other cases, the producers will argue that the cost increases resulting from regulation overstate the external costs, or that the external costs are transitory in nature and these short run effects should be ignored.

In view of a recent forecast of the tripling of U.S. coal production to 1.5 billion tons by 1985,

most of it produced by strip mining methods, policy
makers need to develop adequate evaluation techniques
that can:

- identify the sources of externalities
- place a valuation on the external costs
- develop a notion of the time effects on
  these costs
- monitor the effects of policy changes in
  abating undesirable externalities.

In this paper, the feasibility of using local
property tax assessment mechanisms as a major source
of evaluating the external effects of strip mining
will be examined. In addition, the possibility of
developing a mechanism for monitoring the effects of
regulation will be studied. The property assessment
mechanism has the inherent appeal of using detailed
local data that is already being produced in every
state as part of its administration of property
taxes, despite the varying degree of consistency and
accuracy. This, consequently, could be both a cheap
and effective tool for evaluation.

The preliminary development of this mechanism
has consisted of the building of a model in which
the relationship between strip mining and local
property tax assessment is defined. Then, the
methodology and data needed to test the model for
Ohio strip coal mining will be discussed. Finally,
the results of several empirical tests and conclusions and recommendations resulting from these tests
will be presented.

In the strip mining industry, externalities have
been noted as stemming from acid water run-off,
siltation and sedimentation of ponds and lakes,
destruction of land cover, and a decrease in the
aesthetic value of the landscape.[1] All of these
effects are held to have caused losses to owners of
adjoining property in the form of bad water, increased
danger of flooding downstream, and landscape vistas
like the backside of the moon.

Howard's[1] attempts to estimate the costs of these
types of externalities in the Kentucky coal fields
relied on an inordinate amount of scientific, wild-
eyed guessing (SWEG) type data simply because there
was no convenient way to estimate the costs of many
of the effects. An even greater deficiency in the
Howard estimates stems from the lack of any provision
for gauging the effects of legislation on externalities
over time.

Another external cost not dealt with by Howard was the destruction of the local property tax base that was forecast in the post World War II days of the strip mining industry.[2]   Indeed, this type of loss was deemed one of the most important effects of strip mining by Caudill[3] who wrote:

> The long-range impact of such wrecking (surface mining) on the economy of an already poor and backward state, is incalculable. While mountain land is now assessed at very low values, strip mining often eliminates it from taxation entirely. Thenceforward, the mountaineers who own the surface regard it as worthless for all purposes and decline to pay any taxes on it. If the state goes to the expense and effort required to sell the land at a tax sale, there are no buyers. Thus the region's schools and other public facilities are deprived of desperately needed revenues, and the taxpayers of more prosperous areas of the state are compelled to produce new funds for the support of the "pauper counties" in the plateau.

By contrast, the coal industry has painted the disruption caused by strip mining as a temporary condition that is cured by reclamation programs of responsible coal operators and contends that the land is restored to "beneficial surface use." The National Coal Association,[4] after citing examples of reclamation projects, claimed "still other reclaimed areas are providing attractive sites for private homes, churches, schools, theaters, shopping centers, and industrial parks."

The importance of attempting to verify these claims and counter claims rests not on declaring a winner in this debate but in formulating a workable public policy that will allow the coal industry to satisfy energy needs of the nation without wrecking the environment. A wise policy will balance the potential benefits against losses so that accurate estimates of external costs are essential.

It must also be noted that the distributional effects of benefits and costs must be included in the policy-making decision. Caudill contends that the benefits of cheap energy are being enjoyed outside of the coal-producing regions at great cost to those that remain in those regions.

We will attempt to make a preliminary assessment of the external effects of stripping following the suggestions of Graham and Caudill[2,3] that the external

effects of strip mining can be measured in terms of
their effect on local property tax assessment since
these effects are largely local in nature.

Undesirable side effects of strip mining would
show up in the form of lower property values as land
is strip mined. Only acid water run-off damage, in
the form of caking in boilers and corrosion of boats
and bridges at considerable distances downstream,
and floodplain damage would not show up in market
valuation of adjacent properties.[5]

While each of the separate interests in land,
such as surface and mineral rights, are theoretically
subject to the general property tax,[6] in practice
mineral rights are seldom taxed because assessors
do not have the ability to evaluate geological
information.

The complaint by Caudill about strip mining
destroying the tax base hints tnat the extent of the
external effects can be measured by its effect on
the real property tax base. The word "real" comes
from the Latin word *res* and translates as "thing."
Real property, the most enduring and least portable
commodity, has been viewed as being immovable and
fixed in its location. Generally included as real
estate are growing crops, standing timber, and
minerals in place.

A testable theoretical hypothesis would be that
local property values would either decrease in value
or grow at a slower rate in an area where strip
mining has taken place than in an otherwise similar
nonmining area.

Since the mining activity may induce a temporary
increase in real property values as a result of
operations such as the construction of coal tipples,
we would add that the property value response would
be a lagged one. Another reason to stipulate a
lagged relationship would be the fact that assessments would start to fall only in the next assessment
period after mining had taken place.

If strip coal mining areas show growth rates in
tax valuation similar in sign and strength to growth
rates in nonmining areas, the hypothesis would have
to be rejected. If growth rates in strip coal mining
areas in some latter period exceeded those in nonmining areas, we might be persuaded to accept the
contention of the National Coal Association that
stripping actually holds long term benefits for an
area.

It is also conceivable that the selection of a
suitable time period for lagging the amount of coal
mined might turn up a situation where the initial

hypothesis might be accepted when using some period, t+1, but rejected in some longer run time period, t+2.

In the following section, in which we have attempted to make some preliminary empirical tests of the hypothesis, we have avoided this problem by arbitrarily setting the length of time of the lagged response.

THE DATA BASE

The decision to test the above hypothesis using Ohio data was made for a number of reasons: Ohio has been one of the major coal-producing states over the past 20 years; the state has also been following the industry trend of producing an increasing amount of its coal by surface mining. In 1971, surface mining produced about 75% of the 50 million tons mined.[7] The authors had some familiarity with the institutional peculiarities of Ohio property tax assessment practices, and part of the needed data was already available on a machine readable file.

We selected Ohio counties as the initial unit of analysis not because we had any strong feeling that this unit was more appropriate than township ranges, but solely on the ground that the most readily available data was in terms of counties. We are aware that the grain of any land use greatly influences the results. Meaningful economic analysis on any study tied to land is always complicated by the fact that the boundaries of the reporting jurisdictions—in this case Ohio—were largely determined by historical accidents.

Some attempt was also made to aggregate data for comparative purposes using State Economic Areas (SEA's), which are subdivisions of states, consisting of single counties or groups of counties that have similar social and economic characteristics.[8] There are eight SEA's outside of the major urban areas. Each major urban area is treated as an individual SEA.

Ohio property is valued by law at fair market value by county tax assessors. In each of the 88 Ohio counties, assessment proceeds under supervision of the County Tax Division, State of Ohio, using State Tax Bureau rules.

In years past, after the assessment was made, various counties put different percentages of the market valuation on the books as the tax valuation.

This means that the assessed valuation roll consistently understates the true market valuation. However, it does so in a manner that will vary from county to county and between different types of property within some counties.

In a 1972 court ruling, all Ohio counties were ordered to assess all types of property at the same percentage of market value. Since this type of uniformity did not exist in the data for the period we were interested in, it had to be normalized so that cross-county comparisons could be made.

Another problem in making such comparisons stems from the fact that tax assessments are updated in two ways:

1. the annual update that comes from adding the value of construction permits
2. the reassessment of all property in the county, which takes place only once every six years.

The latter type of assessment change is done on a rotating schedule basis so that cross county comparisons could be seriously affected by the year of reassessment in each county.

Fortunately, the County Tax Division maintains an annual county sales ratio correction factor, which we applied to normalize the data. This sales ratio is the quotient of the assessed value of all property transferred within the county for the year, divided by the sales price.

As the instrumental variable to measure changes in county tax valuations for the 1962-1970 period, we calculated $Beta_1$ coefficients for each Ohio county using normalized annual real property valuations as the observations in an ordinary least square regression for each county, so that where t = the year, RPV = the corrected real property valuation, and B_1 = the coefficient in the form

$$RPV = \alpha + B_1 t + e_1.$$

The results had low standard errors and high F values in almost every county. Thus, we believe this to be a statistically significant measure of growth rate, designating the average annual rate of change in real property value for the period 1962-1970. We selected these years because we had these observations available in a machine-readable file and they contained a sufficient number of observations to construct a meaningful trend for property valuation changes in each county.

In order to measure the effects of strip mining, we constructed a variable that summed the tonnage of both strip- and auger-mined coal for the 1955-1962 period[7] in each county and then divided it by the number of square miles of land area in a given county. This variable, which would give an approximate measure of disturbed land in relation to the total land area on the county tax rolls, was labeled the Surface Mined Coal Area Ratio (SMC).

EMPIRICAL RESULTS

Our first level of analysis was performed by separating annual rate of change in property valuation into coal mining counties, those having SMC values greater than 0 and noncoal mining counties, those having an SMC equal to 0. This differentiation was made for the purpose of testing the hypothesis that coal-mining counties have lower property value growth than other counties.

There was a significant difference (at the 99th percentile level) between the growth rate in the 63 noncoal counties (mean = 8.86%) and the 26 mining counties (mean = 5.95%). A very naive interpretation might hold that the difference in the two groups' rate of growth would be attributable to mining. Not so! Since most coal is mined in the predominantly rural eastern and southeastern counties of Ohio, a more probable explanation for the difference in growth rates was the inclusion of the large urban growth rates in the nonmining group.

Our examination of the rural SEA's (mean = 6.44%) and the urban SEA's (mean = 8.62%) revealed that large differences were attributable to the rural or urban nature of a county. However, the coal county growth rates (5.95%) were significantly lower than the growth rate of all rural counties (6.44%). If the 1962 normalized tax base of the 25 coal counties was multiplied for the nine year period by the annual growth rate differential between the two groups ($7,458,404,760 X 9 X 0.0049 = $328,915,650), an estimated value of the costs of the growth rate would exceed $328 million in property valuation.

It might be argued that comparison of coal counties to all rural counties does not take account of other regional differences, and that attributing the differential to coal mining was improper specification of causality. We ran an analysis of variance on the four rural SEA's having counties with SMC greater than 0 and found that the coal mining counties within each SEA had lower mean growth rates than nonmining counties in the same SEA. Due to the low number of observations, resulting in few degrees of freedom, the F values in these Anova's caused us to reject the hypothesis that coal and noncoal means within their respective SEA's were significantly different.

A comparison was then made of the rate of growth in SEA-7 and SEA-8. SEA-7 contains five western Ohio River valley counties including Adams County, which has the lowest median family income in the state. Except for the fact that none of these counties mine coal, they might be viewed as having characteristics that approximate those of the Appalachian counties further east in SEA-8. Nine of the eleven SEA-8 counties mine coal.

The mean rate of growth for SEA-7 (mean = 5.36%) was 0.92% higher than that of SEA-8 (mean = 4.44%). Even if the westernmost, Clarmont County, is excluded from SEA-7 on the grounds that its property values are affected by the growth of urban Cincinnati, the difference amounts to 0.71%. These findings tend to be even greater than the earlier rural coal and noncoal differentials.

As the last part of our preliminary testing of the effects of strip coal mining, we set up a regression model to test the effects of the SCM variable on the $Beta_1$ values for mining counties in the form

$$B_1 = \alpha + B_2 SCM + e.$$

The relationship between B_1 and B_2 was expected to be negative, indicating that the more coal mined in relation to land area, the lower annual rate of growth. In the calculated regression,

$$B_1 = 11.630 + .047 SCM \quad (0.134 \text{ Std. Error})$$

not only did the coefficient of SCM have the wrong sign, but it was statistically insignificant. We therefore had to reject the hypothesis that the size of the SCM variable explained differences in annual growth rates in the coal mining counties.

CONCLUSIONS

Our findings support the thesis that strip coal mining counties in Ohio have lower growth rates in terms of property valuation when compared with other rural counties. While the tax base has not declined in value, it has failed to appreciate at a rate comparable to other counties. If the 0.49% rate is accepted as approximating the differential in growth, we can state that the annual cost over the 1962-70 period was in excess of $36 million in lost property values. However, the failure of the SCM variable to explain the difference suggests that this loss might be attributable to general regional conditions rather than to the effect of strip mining alone.

It may well be that the introduction of the SCM variable differential, with respect to time, would better explain differences in Beta$_1$'s for the coal mining counties. Our comparison of SEA-7 with SEA-8 tended to support the 0.49% differential rate and would hold out hope that the development of a more sophisticated regional model could more accurately assess the external effects of stripping.

With the introduction of a more accurate time lag into the model, it may then be possible to monitor the effects of strip mining legislation. For example, if we found that the mining county differential tended to narrow in the period after the enforcement of the 1972 legislation began, benefit-cost studies could be undertaken on an annual basis. As in other econometric studies of this type, further disaggregation of the data might greatly add to the confidence that could be placed in the results. Both coal mining and tax assessment data could be obtained down at the township range level, albeit at a greater cost.

REFERENCES

1. Howard, H. A. "A Measurement of the External Diseconomies Associated with Bituminous Coal Surface Mining, Eastern Kentucky, 1962-1967," *J. Nat. Res.*, *11*, 148 (January, 1971).

2. Graham, H. D. "The Economics of Strip Coal Mining," *University of Illinois (Urbana) Bulletin*, *46(17)*, 89 (1948).

3. Caudill, H. M. *Night Comes to the Cumberland.* (Boston: Little, Brown & Company, 1963), p. 317.

4. National Coal Association. *Bituminous Coal Facts, 1966,* (Washington, D.C.: National Coal Association, 1966), p. 29.

5. Brooks, D. B. "Strip Mine Reclamation and Economic Analysis," *Natural Resources Journal, 6(1),* 126 (January 1966).

6. Cooley, T. M. *The Law of Taxation.* (Chicago: Callaghan and Company, 1924), p. 1218.

7. Ohio Department of Industrial Relations. *Division of Mines Report,* Columbus, Ohio (annually).

8. U.S. Department of Commerce, Bureau of the Census. *1970 Census of Population, State Economic Areas.* Washington, D.C. (1972), p. iv.

22. DECENTRALIZATION OF URBAN ECONOMIC ACTIVITY; ITS IMPACT ON ENVIRONMENTAL MANAGEMENT

Robert M. Clark
Office of Program Coordination,
National Environmental Research Center,
U.S. Environmental Protection Agency
Cincinnati, Ohio

Haynes C. Goddard
Solid and Hazardous Waste Research Laboratory,
National Environmental Research Center,
U.S. Environmental Protection Agency
Cincinnati, Ohio

ABSTRACT

There is a continuing trend for more and more Americans to live and work in an urban environment. In conjunction with the trend away from rural life is a trend toward suburbanization and the spread of urban activity over larger areas. This spreading of urban activities has many implications for environmental management.

This paper discusses the factors that are causing urban decentralization and examines some of the data that describe this phenomenon. The impact of this decentralization trend upon the kinds of technology that must be used to manage environmental problems is considered. Solid waste management is used as an example.

INTRODUCTION

The increasing urbanization of the United States is well-documented. In 1960, about 70% of the U.S. population resided in areas classified as urban by

the Census Bureau. In 1970, this figure had risen
to 73%, and it is estimated that it will reach 79%
by 1980.[1] These statistics demonstrate the importance
of urban areas in the United States and emphasize the
continuing trend for more and more Americans to live
and work in an urban environment. In conjunction
with the trend away from rural life, which would
seem to imply the growth of large central cities,
is a trend toward suburbanization and the spread of
urban activities over larger areas. Since many environmental problems are associated with urban life,
this spreading or decentralization of urban activity
is making environmental management increasingly difficult and costly. The spread of urban activity
results in decreasing densities of residential and
commercial land use and increases the area over which
production and consumption residuals must be managed.
For example, water and wastewater treatment systems
have significant economies of scale associated with
high density usage. Mobile source emission from
automobiles and stationary source emission from
manufacturing plants become much more difficult
management problems when they come from decentralized
sources. Increasing solid waste generation from
diverse sources raises the costs of collection and
disposal systems by increasing transportation costs,
creating scarcities of nearby land for disposal, and
limiting the number of options available to the
solid waste manager.[1]

Based on this brief argument, it is possible to
hypothesize that the decentralization of urban
activity causes increases in the costs and problems
associated with environmental control and management
in general. The purpose of this paper is to analyze
those factors that relate the increasing costs of
urban solid waste management, in particular, to
economic decentralization. In addition, the available empirical evidence describing the decentralization phenomenon will be examined. Some appropriate
guidelines will also be given for choosing technology
and strategies to control the solid waste problem.

URBAN AREAS

To properly understand the decentralization process and its environmental impacts, it is important
to understand some of the factors that lead to the
existence of cities and to their continuing
decentralization.

The Structure of Urban Complexes

The large size and rapid growth of urban areas have frequently been observed to be responses to the income and employment opportunities that such areas provide.[2] The nature of the goods and services produced in urban areas can, therefore, be assumed to be different from those produced in rural areas. Significant differences between urban areas of different sizes can also be assumed. In addition to these differences in production characteristics, there are variations in the nature and intensity of land use from city to city and from one part of a city to another. Skyscrapers and high-rise apartments dominate the center of a city, but toward the outskirts, single-story factories and single-family homes prevail. This phenomenon may be characterized as factor substitution, in which structures or capital are substituted for land at locations closer to the central city. Land costs generally increase as one approaches the central city. This substitution effect causes a tremendous difference in capital/land ratios and is the market's response to the greater scarcity of land nearer the center of cities.

There are a number of reasons why various activities tend to be located close together in cities. The most important of these factors are the scale economies that occur when competing firms or firms who depend on each other's services are located to take advantage of their proximity. In addition, there are many specialized businesses and consumer services for which the per-business or per-capita demand is so small that a large city is needed to support even a few suppliers.

A logical conclusion is that the importance of central location to efficient production of goods and services, or to increasing returns to scale, will insure conditions that favor the creation and growth of cities. On the other hand, conditions of production probably impose a definite limit on the efficient size of cities. Consider, for example, the possibility of doubling the population of a city by doubling the height of every building. If this were feasible and if twice as many people were to travel between each pair of points, the demand for transportation would be doubled. But if transportation required land as an input, more land would be used after the doubling of population than before. Thus, some land previously used for buildings would then be used for transportation, thus requiring new buildings at the edge of the city. By thus moving the edge of the city farther

out, some people would make longer trips than before,
and more transportation inputs would be required.
Thus, a doubling of the city's population requires
more than doubling transportation inputs. A city
will eventually reach a size in which this lack of
economy in transportation will offset the economy of
size resulting from increasing returns in production.

The decentralization or suburbanization of
economic activity in major metropolitan areas, particularly those that grew to immense size during the
19th century, is a familiar phenomenon. Associated
problems such as increasing difficulties with city
finances are also familiar.

The Decentralization Process

Large urban areas in the United States have re-
curring and fairly predictable land use patterns.
Central cities tend to specialize in manufacturing
and service trades, and areas more distant from the
central city tend to be residential, with manufacturing and retail centers at the suburban fringe.
These spatial trends are the result of a single basic
economic institution: the urban land market. The
competition for land, determined by one's ability to
pay and one's preferences, determines this pattern.
Thus, the density of production and consumption
activities is a function of land prices and the possibility of substituting land for capital. Because
of rising transport costs, land prices decrease as
the distance from the central city increases.

In recent years, however, there has been a great
deal of investment in high-speed, high-volume highways
in and around urban areas. The investment has reduced
the cost of transportation to the users of these
systems. These falling costs have probably been the
single most important factor leading to decentralization. A second factor is rising levels of per capita
income that permits consumers to exercise their
preferences for the amenities of lower density living.

On the assumption that some central district has
been established for every city, economists have been
able to construct models of the demand for land based
on the advantages of proximity to that center.
Business firms and households must trade off the
advantages of proximity to the center of the city
against the cheapness of land farther away.

Mills[3] has developed a schematic diagram that
classifies urban activities into four sectors and
depicts the rent gradient for each (Figure 48). In

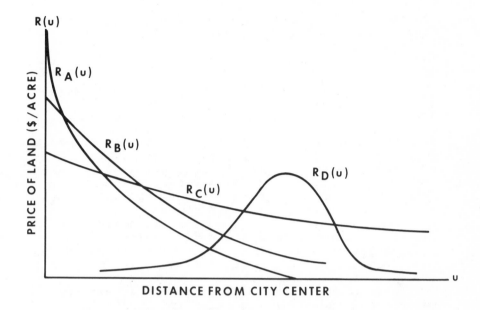

*Figure 48.   Theoretical analysis of urban structure. Sector A
is nearest the center of the city and represents
centrally-located manufacturing and office activities.
Sector B is the second nearest to the center and
represents low-income housing. Sector C is made up
of suburban, high-income housing, and Sector D is
suburban manufacturing and office activities.*

this case, rent is the amount of money an individual
or firm is willing to pay for a unit of land. The
four sectors represent centrally located manufacturing
and office activities, low-income housing near the
center of the city, suburban high-income housing, and
suburban manufacturing and office activities.   The
latter sector is assumed to be able to pay more rent
for suburban land than for a central city location
since intercity highway shipments of freight are
cheaper from suburban areas.

Cheaper land may induce a family to consume more
living space at the expense of other items of consumption and increased commuting time and costs.
The choice of more space can be more easily exercised
at higher incomes than at lower ones; consequently,
a model based on land value gradients will predict
that under some circumstances, more affluent families
will live farther from the city center than those of

limited means. Low-income families will tend to
live on smaller plots of high-rent land.

Businesses, too, will vary in their choice of
location. Some small manufacturing industries and
office activities require rapid access to each other,
so they must choose central location over cheapness
of space. Offices are typically forced upward into
skyscrapers as central space becomes more costly.
On the other hand, some manufacturing industries
need abundant land, and as the cost of land increases,
they tend to move farther from the city. Major retail establishments where infrequent purchases are
made by each buyer tend to seek central locations,
where the population of the entire city will have
convenient access. This same pattern occurs with
some cultural centers. Other more standardized
retail and service activities (including, recently,
some larger ones) in which per capita purchases are
more frequent have located in residential districts
closer to their customers.

There are important similarities between the
theories of industrial location and household location. Firms are assumed to maximize profits (or at
least minimize costs) by choosing a location for
production and then shipping the commodity they produce to the center of the city. Households are
assumed to maximize their utility or satisfaction
by choosing a residential location (among other goods
and services) and then "shipping" the worker himself
to the center of the city. Thus, production of
housing services is analogous to industrial production, and commuting is analogous to the shipment of
commodities.

All evidence indicates that decentralization of
both residential and manufacturing activities is
continuing. Further improvements in transport technology and increasing highway investments provide
strong impetus for this, as do the increases in per
capita income that allow individuals to exercise
their preferences for reduced-density living. These
trends have important implications for those environmental management functions such as solid waste
management, whose costs are a function of households
and industries. But before analyzing these implications, it is important to review some of the empirical
evidence regarding the decentralization process.

EMPIRICAL ANALYSES OF DECENTRALIZATION

Static Analysis

The statistics discussed here document the trend
toward decentralization of urban areas. The differences between central city and suburban populations
are examined using a representative sample of Standard
Metropolitan Statistical Areas (SMSA's). The SMSA
was chosen as the basic unit of analysis for this
discussion because it consists of one central city
with a population of at least 50,000 and of surrounding counties that are metropolitan in character.
SMSA statistics, therefore, do not require data from
subdivisions smaller than counties, and the statistics are relatively plentiful. The employment and
residential distribution for 90 relatively self-
contained SMSA's have been analyzed, and the data
are shown in Table XXIX.[4]

All 90 SMSA's represented in Table XXIX had a
population of at least 250,000 by 1947. This table
shows that population was more suburbanized in
1960-63 than employment was, since 60% of the population lived in the suburbs, and 49.6% of the population worked in the suburbs. However on a percentage basis, employment suburbanized at a faster
rate than population since in 1947-50, 47% of the
population lived in the suburbs, and 32.7% of the
population worked in the suburbs.

Dynamic Analysis

The data analyzed previously were taken from
census figures showing percentages of people living
and working in central cities and suburban rings.
Although such data can reveal the broad outlines of
the massive postwar movement to the suburbs, they
are subject to severe limitations.

Several studies have provided strong evidence
that population density falls off fairly smoothly
and at a decreasing rate as one moves out from the
city center.[4] These studies have found that the
negative exponential function provides a good
approximation to these decreasing population
densities. It can be written:

$$D(u) = De^{-\lambda u} \qquad (1)$$

Table XXIX

Suburbanization of Population and Employment in
90 Standard Metropolitan Statistical Areas, 1947-63

(In thousands)

Sector	1947-50[b]				1960-63[c]			
	In Central City		In Suburban Ring		In Central City		In Suburban Ring	
	Number	Per Cent	Number	Per Cent	Number	Per Cent	Number	Per Cent
Population	26,742	53.0	23,506	47.0	26,554	40.0	39,483	60.0
Employment								
Manufacturing	3,750	60.5	2,449	39.5	3,250	46.2	3,791	53.8
Retailing	2,032	71.5	811	28.5	1,667	48.7	1,756	51.3
Service	673	79.5	173	20.5	826	61.1	525	38.9
Wholesaling	980	85.2	171	14.8	943	65.2	503	34.8
Total Employment[a]	7,435	67.3	3,604	32.7	6,686	50.4	6,575	49.6

[a] Figures may not add because of rounding.

[b] Source: Population data from U.S. Census of Population, 1950; employment data from U.S. Census of Business, 1948, and U.S. Census of Manufacturers, 1948.

[c] Source: Population data from U.S. Census of Population, 1960; employment data from U.S. Census of Business, 1963, and U.S. Census of Manufacturers, 1963.

where D(u) is the density u miles from the center,
e is the base of the natural logarithm, and D and λ
are parameters to be estimated from the data. D is
the measure of density at the city center, and λ,
which is positive, is a measure of the rate at which
density declines with distance from the center. If
λ is large, then it follows that the density falls
off rapidly; if it is small, density falls off slowly.
This density function allows us to make comparisons
across space and time that a static analysis will
not allow. To illustrate the use of this type of
analysis, population density functions were calculated for 18 SMSA's for 1948, 1954, 1958, and 1963
(Table XXX).[4] Calculated density functions display
a remarkably consistent pattern of flattening
through time. Albuquerque is the only SMSA that
shows increases.

Figure 49 illustrates this decreasing density
effect for Columbus, Ohio. The analysis is performed
from the city center at 1-mile increments over a
10-mile distance. A significant difference can be
seen in population densities over time because of
the decentralization effect.

Figure 50 shows the tail of the population distribution curves illustrated in Figure 49. Assuming
a rural population density of 250 persons per square
mile, we can see that this frontier has moved from
approximately 6.5 miles in 1948, to approximately
11 miles in 1970. The impact of this effect on
available land for solid waste disposal is that
potential sites (as represented by low population
densities) must be located farther and farther from
the city center. Correspondingly, transportation
costs for solid waste management must increase
because of this decentralization effect.

Given consumer preferences for low density living
and assuming that per capita income will continue to
increase, it can be reasonably expected that decentralization will continue and that the cost of
maintaining a constant level of output for solid
waste management services will increase. The basis
for such a conclusion is discussed in the following
section.

Decentralization and the Cost
of Solid Waste Management

The increased cost of solid waste management,
as a result of urban decentralization, involves one
main factor: the dependence of collection, haul,
and disposal costs on population density.

Table XXX

*Density Functions for Eighteen Standard Metropolitan
Statistical Areas in the United States*

Metropolitan Area		Population 1948	1954	1958	1963
Albuquerque	λ	.56	.61	.61	.62
	D	5,748	11,387	14,148	18,180
Baltimore	λ	.48	.40	.36	.33
	D	51,159	42,693	37,481	34,541
Boston	λ	.27	.25	.23	.21
	D	35,473	32,629	28,630	24,922
Canton	λ	.69	.62	.58	.54
	D	19,994	18,724	17,610	16,591
Columbus	λ	.78	.65	.58	.52
	D	44,303	38,680	34,643	31,710
Denver	λ	.59	.45	.38	.33
	D	27,779	22,884	19,678	18,008
Houston	λ	.37	.28	.24	.21
	D	15,156	13,118	11,881	11,243
Milwaukee	λ	.47	.37	.32	.27
	D	58,318	44,262	37,823	31,123
Philadelphia	λ	.31	.27	.25	.23
	D	53,264	45,714	41,868	38,268
Phoenix	λ	.51	.39	.33	.28
	D	11,324	11,244	10,350	9,521
Pittsburgh	λ	.27	.25	.24	.22
	D	25,072	22,780	21,699	18,974
Rochester	λ	.73	.55	.47	.40
	D	39,682	28,194	24,033	20,527
Sacramento	λ	.77	.56	.48	.41
	D	22,120	18,337	16,782	15,262
San Antonio	λ	.63	.56	.50	.45
	D	27,513	28,705	25,855	23,951
San Diego	λ	.27	.23	.21	.20
	D	10,438	12,583	13,164	14,972
Toledo	λ	.83	.72	.67	.61
	D	41,123	34,661	31,768	28,151
Tulsa	λ	.89	.63	.50	.40
	D	28,788	20,126	15,339	11,947
Wichita	λ	.98	.74	.63	.54
	D	29,149	23,589	20,153	17,613

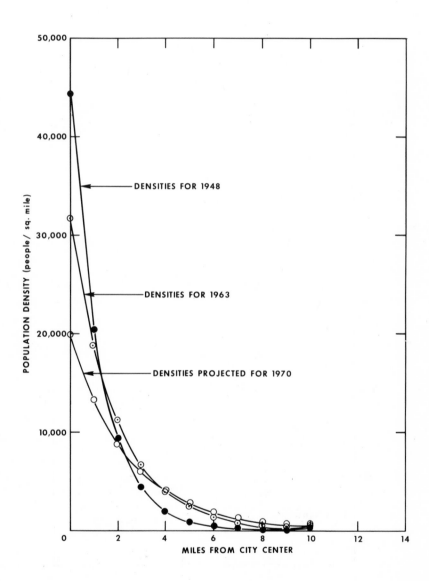

Figure 49. Population density patterns for Columbus, Ohio.

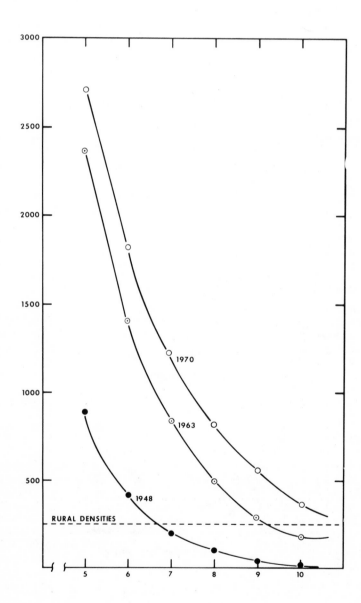

Figure 50.   Tail of population density curves for Columbus, Ohio.

Collection

Solid waste management services must be produced at spatially separate waste generation points. Decentralization causes the distance between these points to increase over time, resulting in greater time and fuel costs. A simple but formal causal model would contain the following variables and relationships:

$$y = f_1(t), \quad f_1 > 0 \qquad (2)$$

where income grows over time (y = income; t = time);

$$d = f_2(y,k), \quad f_y' > 0, \quad f_k' > 0 \qquad (3)$$

where distance (d) from central business district increases with y income and highway investments (k);

$$R = f_3(d), \quad f_3' < 0 \qquad (4)$$

where land rent (R) falls with distance from the central business district;

$$D = f_4(R), \quad f_4' > 0 \qquad (5)$$

where density (D) falls as land rents fall;

$$CC = f_5(D), \quad f_5' < 0 \qquad (6)$$

where collection costs rise as density falls.

Applying the chain rule to collection cost with respect to time, $\partial CC / \partial t > 0$, collection costs per household can be expected to increase over time. The other major variable affecting collection cost is highway investments. Highway investments (k in Equation 2) are also related to distance in a positive manner and serve to decrease population density; thus,

$$\frac{\partial CC}{\partial k} > 0.$$

Haul

Given that a community has acquired a large disposal site that has the capacity to accept the waste

load generated over a 20-year time horizon (the
typical practice), then decentralization of economic
activity will increase the haul distances from the
collection routes to the disposal site. Actual time
spent on the route will thus be reduced, along with
productivity, and costs will be expected to rise.

Disposal

Disposal here refers to any final waste disposal
technology that requires long-range planning decisions and that is a permanent fixed investment.
Landfill and incineration are two examples.

As decentralization progresses, there is an increased likelihood that the economic life of large
fixed investments will become shorter than the planned
physical life; that is, planning will err on the side
of over-capacity and longer-than-needed capital
durabilities.

Given that demand for disposal depends on household waste generation, decreasing densities and
increased suburbanization result in reduced demand
for a fixed facility planned on the basis of past
densities. Where scale economies are important,
operating costs will increase in addition to the
capital costs discussed above. This fact is illustrated in Figure 51, where $Q^d = f(D_1)$ is quantity

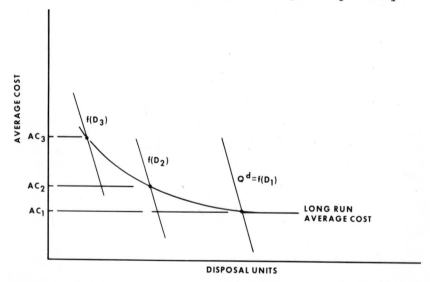

Figure 51. *Quantity of disposal demanded as a function of
population density ($Q^d = f(D_1)$).*

demanded as a function of population density in period one. Declining density will reduce the demand for a given facility to $f(D_2)$ and $f(D_3)$, resulting in an increase in operating costs.

There are essentially two ways in which the demand for a disposal site can fall off—through incorrect planning and increased haul costs. An estimate of future waste generation is needed to know which size disposal facility is required. Projections based solely on waste/population ratios will understate future waste generation unless both decentralization and per capita income trends are taken into account.

If a projection of waste generation for a jurisdiction with currently undeveloped land is made on the basis of a constant density coefficient, implicitly or explicitly, then actual generation will fall short of the projection. Thus, the average cost of operating a landfill will rise (barring waste imports from other jurisdictions).

Furthermore, if decreased densities lead to increased marginal haul costs, a point may be reached where the cost of opening a series of dispersed disposal sites may be less than the higher haul costs. This situation will also result in an apparent disparity between the economic and physical lives of the fixed facility. It is also possible that the next best alternative is to maintain the larger initial facility and operate a transfer system. Then the increased costs, resulting from decreased densities, would appear only in the haul activity.

Thus, it is possible to hypothesize that the decentralization of economic activity will affect all three phases of solid waste management. A lack of data currently limits rigorous tests of this proposition, but empirical evidence obtained from various sources can be presented to support this hypothesis.

EMPIRICAL EVIDENCE

Collection Costs

The effect of decreased population densities on collection costs has been shown in some tentative statistical analyses performed for the Solid Waste Research Laboratory of the National Environmental Research Center, Cincinnati.[5] Two regression equations are of interest:

$$Q^d = 3810x_1^{-0.704} x_2^{-0.124} x_3^{-0.26} \quad (R^2 = .6875) \qquad (7)$$

where Q^d = per capita waste generation, x_1 = average revenue (a price proxy), x_2 = per capita income, and x_3 = population density.

The coefficients of x_1 and x_3 are significant at the 0.005 level. This result implies that a 10% decrease in density will lead to a 2.6% increase in waste generation, thus, raising the demand for collection and disposal.

The second equation gives the impact of density changes on solid waste management costs (excluding disposal):

$$TSC = 9.15D_1 + 14.05D_2 - 0.409x_1 + 0.0014x_2 - 0.0034x_3$$

$$(R^2 = 0.6480) \qquad (8)$$

where TSC = total systems costs per capita, D_1 = dummy variable for cities without direct user charges, D_2 = dummy variable for cities with user charges, x_1 = population density, x_2 = per capita income, and x_3 = wage rate for collection crews. This result suggests that total systems costs rise as density falls.

These two equations provide empirical support for the hypothesis that solid waste management costs increase as production and residential activity decentralizes. There are two reasons for the observed cost increases: increased waste generation and increased distances between waste generation points and between collection routes and disposal sites. Both these factors are the results of lower population density, which in itself implies greater distances between generation points and larger lots and structures that generate more yard wastes and increased purchases (with attendant wastes). Greater waste generation will raise total costs and average costs per household, but probably not marginal costs. All costs, however, (total, average, and marginal) are raised by the increased distances between waste generation points.

Haul Costs

For purposes of analysis, haul from the route to the disposal site will be considered a separate function, although it is frequently combined with disposal. The hypothesis that haul costs tend to rise with declining density is supported by data from a case study performed for Cleveland, Ohio.[6]

Haul costs are dependent, of course, on the location of the disposal site. This analysis is conducted on the assumption that a given disposal site remains fixed during a period of time in which decentralization occurs. Since current practices tend to favor fairly large and long-lived sites (20 years), this assumption is fairly realistic.

A simulation analysis was performed on Cleveland's collection operation, assuming two disposal alternatives—an 8-mile round trip from a typical route and a 40-mile round trip from a typical route. Figures 52 and 53 show the results of this analysis for various collection configurations. The ratio of productive time to transport time is twice as large with the short-haul alternative as with the long-haul option. As the population of the area continues to decentralize, the need for longer haul distances will increase, a result that is representative of most metropolitan areas whose population is decentralizing.

Disposal

There has not yet been any documentation of decreasing densities affecting the costs of operating a disposal site, but it is possible to illustrate the general problem with data from Cleveland.

A central incineration system (a large, fixed facility) that would adequately serve the City of Cleveland would require an investment of nearly $50 million. Figure 54 shows the projected total population and associated waste generation for the city during the period 1973-2000. These figures include an assumed 20% increase in per capita waste generation over this time period and a decrease in population resulting from decentralization factors. The current generation is nearly 375,000 tons per year, and it is estimated that the generation will be 325,000 tons in the year 2000, a decline of 13%. An assumed debt burden of $50 million and a design of 20 years gives a fixed annual cost of $5 million

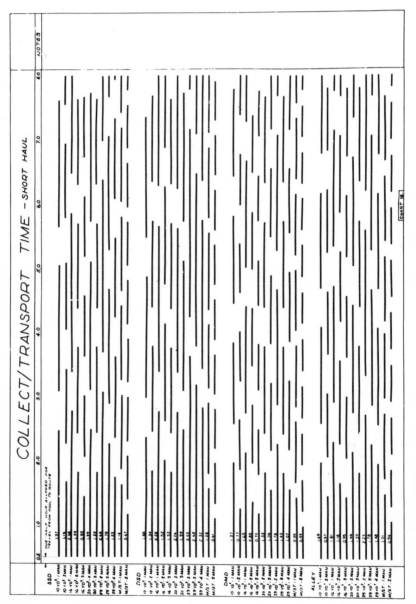

Figure 52. Simulation analysis of a short-haul collection operation in Cleveland, Ohio.

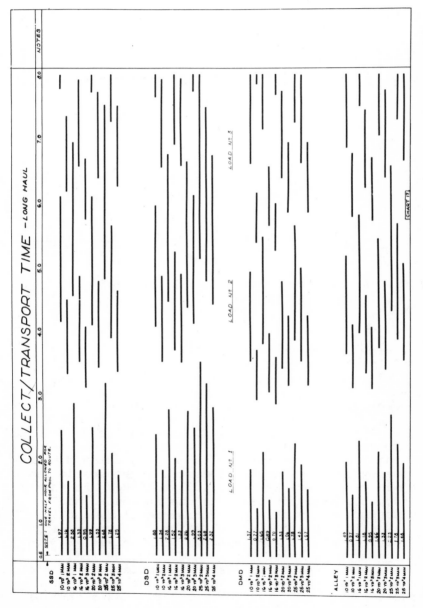

Figure 53. Simulation analysis of a long-haul collection operation in Cleveland, Ohio.

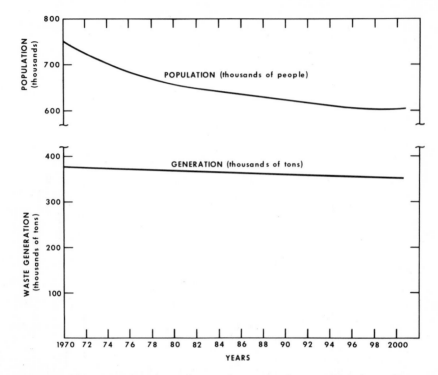

Figure 54. Waste generation and population predictions for Cleveland, Ohio.

per year to amortize the initial capital cost. Dividing this annual cost by the waste generation in 1973 and projected waste generation in 1993, results in costs of $13.35 and $15.40 per ton, respectively, in constant prices (Figure 55).

The principal advantage of a large, relatively capital-intensive facility such as an incinerator is to gain economies of scale for large outputs. In this case, however, projected population trends actually lead to increased costs, increases in per capita waste generation notwithstanding. Furthermore, this analysis neglects the increased haul costs implied by a large central facility and the possibility of altering jurisdiction boundaries to permit importation of wastes. Nonetheless, $13 to $15 per ton is a relatively high-cost system.

A lower-cost option is the transfer system, for which there are basically two alternatives: truck transfer and rail transfer. The fixed and variable

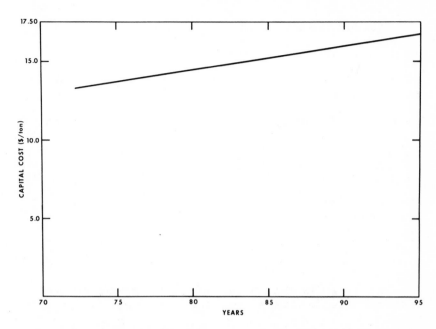

Figure 55. *Projected capital investment costs for incineration in Cleveland, Ohio.*

costs that would be typical for a truck transfer system with two transfer stations assuming a waste generation rate of 260,000 tons/year are as follows:

Fixed Costs:	Dollars
Capital Expenditures:	
Transfer station @ $450,000, amortized for 17 years	45,000
Tractor @ $18,000 for t years	3,780
Special equipment @ $30,000, amortized for 5 years	6,000
Other (Maintenance)	10,000
Labor (Operational Only; Excludes Drivers)	58,000
Fringe (20%)	12,000
Total	134,780

Variable Costs:

Maintenance and Operations	0.40/mile
Vehicles and Drivers	0.39/mile
Total	0.79/mile

From these figures, fixed costs are estimated to be $0.52 per ton for transfer station processing:

$$\frac{\$134,780}{260,000 \text{ tons}} = \$0.52/\text{ton}$$

Variable costs are $0.08 per ton per mile for a 2-way haul:

$$\frac{\$0.79/\text{mile} \times 2}{20 \text{ tons/load}} = \$0.08/\text{ton/mile}$$

The fixed costs for a rail-haul system with 2 transfer stations are listed as follows:

Capital Expenditures:	Dollars
Facilities and equipment, $2,100,000, amortized for 10 years	210,000
Other:	
Utilities	68,000
Maintenance	60,000
Labor	145,600
Fringe (20%)	36,400
Total	520,000

Thus, the fixed costs are approximately $2.00 per ton:

$$\frac{\$520,000}{260,000 \text{ tons}} = \$2.00/\text{ton (bale processing)}$$

Variable costs per ton for a unit train carrying 3,000 tons of waste range from $2.16 per ton to $2.88 per ton, depending on the distance hauled (see Table XXXI).

Both truck and rail transfer systems use land disposal as part of the total collection and disposal system. Land disposal costs in this area range from

Table XXXI

Variable Rail Haul Costs

Miles Hauled	1,500 Tons Per Train	3,000 Tons Per Train*	4,500 Tons Per Train
10	$2.68	$2.16	$1.97
25	2.85	2.26	2.04
50	3.12	2.42	2.17
75	3.39	2.57	2.29
100	3.65	2.73	2.42
125	--	2.88	--

*These values used for sample calculation.

$3.85 per ton near the city to $1.65 per ton 100 miles away. The variable costs associated with these activities are shown in Figure 56.

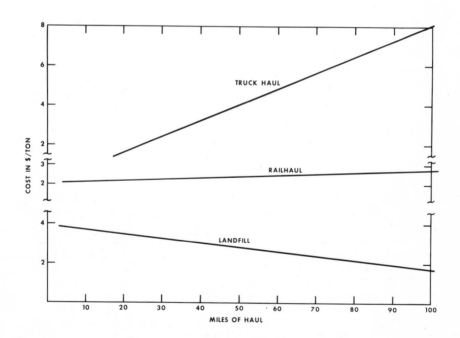

Figure 56. *Variable costs for truck haul, rail haul, and landfill.*

Figure 57 summarizes the total cost per ton for each of the two transfer systems, including land disposal costs. At a haul distance of 50 miles, rail haul and truck transfer reach a break-even point at approximately $7.00 per ton.

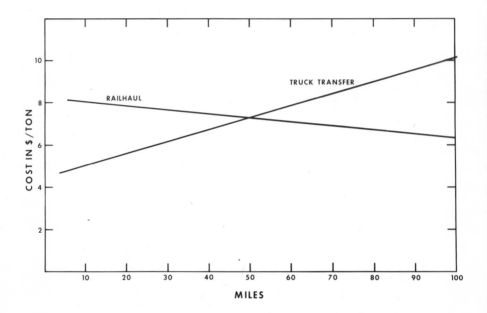

Figure 57.   Total cost for rail haul and truck transfer (including land disposal costs).

The projected land use for Cleveland and Cuyahoga County illustrates the importance of this analysis. Figure 58 shows the generalized present land use in which there is currently some undeveloped land that might be used for land disposal. However, the projected land use for the year 2000 in Cuyahoga County (Figure 59) indicates that there will be no land available for disposal within the county area. This analysis, therefore, suggests the need for a careful examination of new technologies that will recognize the impact of population decentralization.

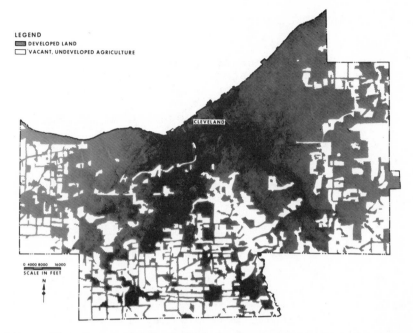

Figure 58. Generalized present land use in Cuyahoga County, Ohio.

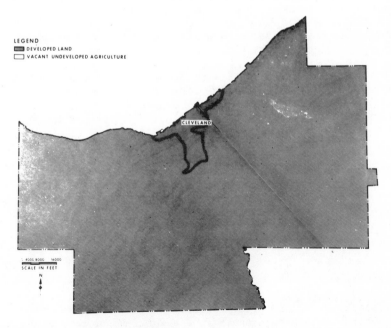

Figure 59. Generalized future land use in Cuyahoga County, Ohio.

SUMMARY AND CONCLUSIONS

The continuing trend toward urban decentralization and the resulting decrease in population densities have been well-documented. The effect of this trend on solid waste management is to increase the costs of collection, haul, and disposal. There appear to be three basic reasons for these higher costs:

1. increased distances between waste generation points and between collection routes and disposal sites
2. increased waste generation resulting from the larger lots and structures characteristic of low density living
3. reduced demand for large, fixed facilities (such as incinerators) that were planned on the basis of the higher population densities of the past.

To avoid higher than necessary costs for solid waste management in the future, it will be necessary to consider carefully whether the economies of scale achieved by large, centrally-located incinerators and landfills are more than offset by increased haul costs. Efficient facility planning must take into account the dramatic increases in haul costs that will occur as population decentralizes. Thus, effective solid waste planning must functionally incorporate decentralization models into the total analysis.

A related question that deserves study is whether reliance on nonmarginal financial mechanisms (property taxes) for solid waste management fosters decentralization.

ACKNOWLEDGMENT

The authors would like to acknowledge the assistance of James I. Gillean, President, ACT Systems, Orlando, Florida, in the preparation of this paper.

REFERENCES

1. Clark, R. M. and R. O. Toftner. "Land Use Planning for Solid Waste Management," *Public Works Magazine* (March, 1972).

2. Mills, E. S. "An Aggregative Model of Resource Allocation in a Metropolitan Area," in *Readings in Urban Economics*. (New York: Macmillan, 1972).

3. Mills, E. S. *Urban Economics*. (Scott, Foresman & Co., 1972).

4. Mills, E. S. *Studies in the Structure of the Urban Economy*. (Baltimore, Md.: The Johns Hopkins Press for Resources for the Future, 1972).

5. McFarland, J. M., *et al*. "Comprehensive Studies of Solid Waste Management," University of California, Berkeley, California, SERL, Report #72-3.

6. Clark, R. M. and J. I. Gillean. "Systems Simulation and Its Application to Solid Waste Planning; A Case Study," Report of NERC-Cincinnati, U.S. EPA (in progress).

INDEXES

AUTHOR INDEX

Ahmad, M. U. 49,54,56
Allen, F. A. 56
Allport, G. W. 249
Aplan, F. 79,93
Appalachian Regional
  Commission 43,48
Arro, A. 135
Baez, A. P. 95
Baker, D. H., Jr. 77
Barnard, P. 94
Barnes, H. L. 74
Bartolatta, R. 230
Beck, I. V. 75
Beck, J. V. 58,60,74
Beckerdite, D. 19
Berstein, L. 230
Bhappu, R. B. 77,84,93
Björkman, O. 181,183
Black, R. J. 191
Bonem, G. W. 26,27,39
Borman, F. H. 229
Bowlus, F. 94
Braley, S. A. 62,74
Brandeberry, J. 184
Brant, R. A. 52,56
Broecker, W. S. 180,183
Brooks, D. B. 261
Broyer, T. C. 230
Brynner, L. C. 62,75
Carlton, A. B. 230
Carr, M. 135
Caudill, H. M. 254,255,260
Chapman, H. D. 229
Clark, R. M. 234,263
Collier, C. R. 51,52,56
Colmer, A. R. 60,74

Cook, T. D. 202
Cooley, T. M. 261
Conrad, E. T. 201,202
Consolidation Coal
  Company 229
Corbett, D. M. 51,56
Cox, A. B. 87,94
Crouch, L. W. 101
Davis, D. 75
Davis, F. T. 83,93
Dean, K. C. 93
Delchamps, E. W. 62,69,74
Department of Natural
  Resources 51,56,249
Donaldson, E. C. 197,202
Dudenkov, S. V. 87,94
Eddy, G. E. 202
Edwards, A. L. 249
Elzam, O. E. 190,211,229,230
Epstein, E. 230
Federal Water Quality
  Administration 75
Fishbein, M. 249
Fisher, E. 75
Forrester, M. L. 181,183
Frank, Gary 135
Ganim, R. J. 144,169
Garrels, R. H. 75
Garrels, R. M. 183
Gaudin, A. M. 87,94
Gillean, J. I. 288,289
Goddard, H. C. 234,263
Graham, H. D. 254,260
Gregor, B. 144,177,184
Halpern, J. 75
Hayward, H. E. 230

299